Engineer's Guide to the
National Electrical Code®

H. Brooke Stauffer

Executive Director of Standards and Safety
National Electrical Contractors Association
Bethesda, Maryland

National Fire Protection Association
NFPA®

nec®

National Electrical Code

JONES AND BARTLETT PUBLISHERS

Sudbury, Massachusetts

BOSTON   TORONTO   LONDON   SINGAPORE

World Headquarters

Jones and Bartlett Publishers
40 Tall Pine Drive
Sudbury, MA 01776
978-443-5000
info@jbpub.com
www.jbpub.com

Jones and Bartlett Publishers
Canada
6339 Ormindale Way
Mississauga, Ontario L5V 1J2
Canada

Jones and Bartlett Publishers
International
Barb House, Barb Mews
London W6 7PA
United Kingdom

Jones and Bartlett's books and products are available through most bookstores and online booksellers. To contact Jones and Bartlett Publishers directly, call 800-832-0034, fax 978-443-8000, or visit our website www.jbpub.com.

Substantial discounts on bulk quantities of Jones and Bartlett's publications are available to corporations, professional associations, and other qualified organizations. For details and specific discount information, contact the special sales department at Jones and Bartlett via the above contact information or send an email to specialsales@jbpub.com.

Production Credits

Chief Executive Officer: Clayton Jones
Chief Operating Officer: Don W. Jones, Jr.
President, Higher Education and Professional
    Publishing: Robert W. Holland, Jr.
V.P., Sales and Marketing: William J. Kane
V.P., Design and Production: Anne Spencer
V.P., Manufacturing and Inventory Control:
    Therese Connell
Publisher—Public Safety Group:
    Kimberly Brophy
Publisher—Electrical: Charles Durang

Associate Editor: Amanda Brandt
Production Supervisor: Jenny L. Corriveau
Associate Production Editor: Jamie Chase
Director of Marketing: Alisha Weisman
Composition: Shepherd, Inc.
Cover Design: Anne Spencer
Interior Design: Kristin E. Ohlin
Cover Image: © R/ShutterStock, Inc.
Printing and Binding: Malloy, Inc.
Cover Printing: Malloy, Inc.

Library of Congress Cataloging-in-Publication Data
Stauffer, H. Brooke.
    Engineer's guide to the National electrical code / H. Brooke Stauffer;
    National Fire Protection Association.
        p. cm.
    Includes bibliographical references and index.
    ISBN-13: 978-0-7637-4886-9 (alk. paper)
    ISBN-10: 0-7637-4886-2 (alk. paper)
 1. Electric wiring. 2. Electric wiring—Insurance requirements.
 3. Electric wiring—Standards. I. National Fire Protection Association.
National electrical code (2007) II. National Fire Protection
Association. III. Title.
    TK3285.S7022 2007
    621.319′24021873—dc22   2007022553
6048

Printed in the United States of America
11 10 09 08 07   10 9 8 7 6 5 4 3 2 1

Table of Contents

Introduction

The *National Electrical Code (NEC®)* is the world's most widely used safety standard. In the United States, it's adopted for regulatory use in all 50 states, U.S. territories, and more than 42,000 local jurisdictions. The *NEC* is also widely used around the globe.

- It is adopted by several other countries.
- U.S.-based multinational corporations follow the *NEC* in facilities around the world.
- The U.S. military follows the *NEC* in facilities around the world.
- The *NEC* is translated into four languages besides English and used in many countries that don't have their own national wiring rules.

For this reason, electrical engineers responsible for designing, specifying, and maintaining premises wiring systems that follow U.S. practice must have a thorough understanding of the *NEC*. The purpose of this book is to introduce electrical engineering students to the *Code* rules that govern installation of electrical power, communications, and control systems.

What This Book Is

Engineer's Guide to the National Electrical Code® is intended to give engineers an overview of how the *NEC* functions as an important tool in the process of designing and specifying electrical systems. This book explains what the *Code* is intended to do and what it isn't; how the *NEC* is organized; the safety rationales that underlie *Code* rules for installing electrical systems; how the *NEC* is used for regulatory purposes; and why the *Code* isn't (and can't be) an all-purpose manual for electrical engineers.

What This Book Isn't

Engineer's Guide to the National Electrical Code® isn't a technical guide or an instruction manual for designing electrical installations in buildings and similar structures. It doesn't cover basic principles of physics and electrical engineering or applied concepts such as energy-efficient design and value engineering. It doesn't teach electrical engineers how to develop plans or specifications for building electrical systems.

Instead, this book explains how the *NEC* functions as one of many tools and inputs that engineers use to design electrical distribution systems for buildings and similar structures. Although the *Code* is a universal starting point for designing and installing electrical systems, many other industry codes and standards expand on its basic principles, functioning as advanced design guides that help electrical engineers meet customer needs in many ways over and above the critical safety rules of the *NEC*. Design references such as the IEEE Color Book series and other standards are listed in Appendix A.

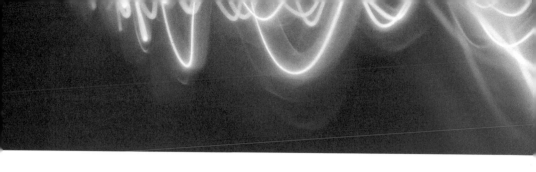

How This Book Is Arranged

Engineer's Guide to the National Electrical Code® doesn't follow the structure of the *NEC*. Instead, it assembles information from many different parts of the *Code* in logical groupings. This is intended to help electrical engineers understand the complexity and interrelationships among different sections of the *NEC* itself and offer engineers a comprehensive view of this global regulatory code.

Engineer's Guide to the National Electrical Code® is a useful introduction to the technical content of the *NEC*. Two other publications provide more detail-level knowledge of actual *Code* rules:

- *User's Guide to the National Electrical Code*® is a tool that helps electrical engineers and designers better understand the purpose, structure, and organization of the *Code*. This illustrated book summarizes *NEC* safety rules in a comprehensive fashion, complete with end-of-chapter exercises to test comprehension.
- The *National Electrical Code Handbook*® contains the complete text of the *Code*, supplemented by explanations and illustrations. It provides a section-by-section, sometimes line-by-line, explanation of key *Code* requirements. The *NEC Handbook* is an essential desk reference for electrical engineers.

How the *Code* Is Made

The *National Electrical Code* is developed under procedures accredited by the American National Standards Institute (ANSI), the federation of U.S. standards-developing organizations. Documents developed under ANSI-approved consensus procedures have the status of "official" U.S. standards. Generally speaking, state and local governments prefer to adopt American National Standards for regulatory use. (NOTE: The terms "ANSI" standard and "American National Standard" are synonymous.)

The consensus process by which the *NEC* and other NFPA documents are developed, including how engineers and other users can participate in helping to create these codes and standards, is described in Unit 7.

H. Brooke Stauffer
Executive Director of Standards and Safety
National Electrical Contractors Association
Bethesda, Maryland

UNIT 1

NEC Structure, Organization, and Language

Introduction

The *National Electrical Code*, ANSI/NFPA standard 70-2008, is one of more than 300 ANSI-approved safety standards published by the National Fire Protection Association (NFPA). It is a model building code intended for regulatory adoption by states and municipalities. This book refers to it interchangeably in four ways: as the *National Electrical Code, NEC, Code,* and (less often) as NFPA 70. This mirrors the way that engineers, installers, inspectors, and other users commonly refer to this important regulatory standard (**Figure 1.1**).

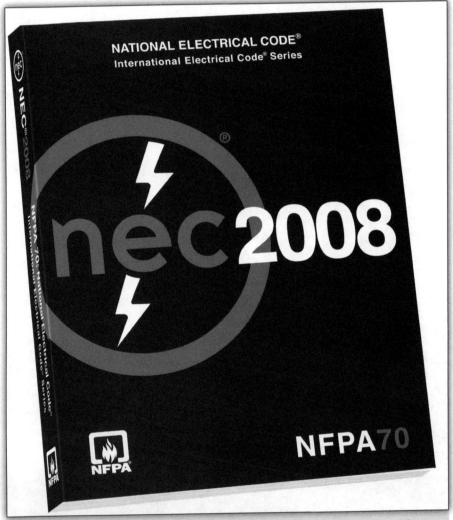

Figure 1.1 The *National Electrical Code* is adopted for regulatory use throughout the United States and around the world. © 2008 NFPA.

This unit discusses the basic structure, organization, and language of the *Code*, as well as its place in the U.S. voluntary standards system. Content, including regulatory and technical concepts that influence how engineers use NFPA 70 as one resource in designing electrical and communications systems for buildings and similar structures, is discussed in subsequent units. This unit covers the following:

- U.S. Standards System
- Terminology Used in This Book
- *NEC* Structure and Organization
- *NEC* Numbering System
- Subdivisions of the *NEC*
- *Code* Language

U.S. Standards System

To a much greater extent than other countries, U.S. commerce and industry depends on a so-called voluntary standards system for regulation of health, safety, performance, and ratings of products and systems. Codes and standards are written by private sector organizations, including trade associations and professional societies, under an open consensus process accredited by the American National Standards Institute (ANSI), and then adopted for regulatory use by federal, state, and local governments.

The U.S. government has recognized the importance of this voluntary standards system for meeting procurement and safety goals through the Office of Management and Budget (OMB) Circular A-119 and the 1996 Technology Transfer Act. Both of these encourage government agencies to depend on private standards organizations to the extent feasible and give guidance for agency participation in this voluntary standards system. Federal participation is coordinated by the National Institute of Standards Technology (NIST), working in close coordination with ANSI.

The success of this approach can be measured by the fact that, over the last decade, thousands of MIL-SPECs (military specifications) and other federal government standards have been canceled and replaced by private-sector codes and standards covering the same products, systems, and processes.

Terminology Used in This Book

The terms "code" and "standard" have particular meanings. NFPA, publisher of the *NEC*, defines them as follows:

> **Code**—A standard that is an extensive compilation of provisions covering broad subject matter or that is suitable for adoption into law independently of other codes and standards. The decision whether to designate a standard as a "code" is based on such factors as the size of the document, its intended use and form of adoption, and whether it contains substantial enforcement and administrative provisions.

Standard—A document, the main text of which contains only mandatory provisions using the word "shall" to indicate requirements, and which is in a form suitable for mandatory reference by another standard or code or adoption into law. Nonmandatory provisions shall be located in an appendix, footnote, or fine print note and are not to be considered a part of the requirements of a standard.

Voluntary Standard

In this book, "voluntary standard" is synonymous with "standard." For example, all documents developed by the NFPA are voluntary standards, though many of them are adopted for regulatory use by state and local governments. By contrast, mandatory standards are developed by government agencies such as the Occupational Safety and Health Administration (OSHA), Environmental Protection Agency (EPA), and Department of Energy (DOE).

Specification, Manual, Instructions, Guidelines

In this book, "specification," "manual," "instructions," and "guidelines" refer to proprietary technical documents not developed through a consensus process. Installation manuals or instructions supplied by electrical manufacturers with their products are examples of these.

Other Terminology and Editorial Practices

Definitions. In general, this book uses terms according to their definitions in *NEC* Article 100. These are discussed in Unit 2.

Rules, requirements, and provisions. *NEC* 90.5 uses the terms Mandatory Rules and Permissive Rules to explain different types of language. However, most *Code* users tend to talk more often about "requirements" or "provisions" of the *NEC*. This book uses the words *rules, requirements,* and *provisions* interchangeably.

Numbers in brackets. Numbers in [brackets] are *NEC* section designations, unless otherwise noted.

NEC Structure and Organization

NEC arrangement is specified in 90.3, and explained here in further detail (**Figure 1.2**).

The *National Electrical Code* consists of an introduction, nine chapters, and eight annexes:
- Article 90—Introduction
- Chapter 1—General
- Chapter 2—Wiring and Protection
- Chapter 3—Wiring Methods and Materials
- Chapter 4—Equipment for General Use
- Chapter 5—Special Occupancies
- Chapter 6—Special Equipment
- Chapter 7—Special Conditions

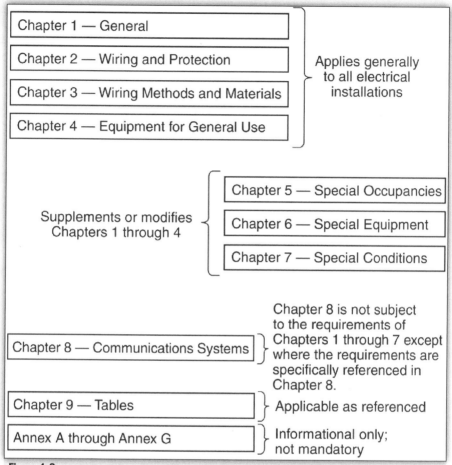

Figure 1.2 Code arrangement. Reprinted with permission from 2005 NEC Handbook, copyright © 2005, National Fire Protection Association.

- Chapter 8—Communications Systems
- Chapter 9—Tables
- Annex A—Product Safety Standards
- Annex B—Application Information for Ampacity Calculation
- Annex C—Conduit and Tubing Fill Tables for Conductors and Fixture Wires of the Same Size
- Annex D—Examples
- Annex E—Types of Construction
- Annex F—Availability and Reliability for Critical Operations Power Systems; Development and Implementation of Functional Performance Tests (FTPs) for Critical Operations Power Systems
- Annex G—Supervisory Control and Data Acquisition (SCADA)
- Annex H—Administration and Enforcement

Introduction (Article 90)

Article 90 doesn't contain enforceable rules, but provides a road map to the *Code* and contains other useful information. This is discussed further in Unit 2: General Requirements.

Chapters 1–4 contain requirements that apply generally to most electric power installations. Subjects covered include definitions; general rules; services, feeders, and branch circuits; load calculations; overcurrent protection; grounding and bonding; conductors, cables, and raceways; and most common types of electrical equipment including switchboards and panelboards, transformers, motors, and luminaires (lighting fixtures).

Chapter 5 covers special occupancies, such as gas stations, aircraft hangars, paint-spraying booths, motion picture studios, and health care facilities. It modifies the general rules of Chapters 1–4 for occupancies that present particular electrical safety concerns. Example: Ignitible dust and fibers found at some industrial plants can be ignited by sparks or heat, so wiring rules for hazardous (classified) locations in Articles 500–504 are designed to prevent this.

Chapter 6 applies to special equipment such as electric signs, cranes and hoists, X-ray equipment, industrial machinery, and swimming pools. It modifies the general rules of Chapters 1–4 for these types of equipment that present particular safety concerns. Example: Because combining water and electricity heightens the risk of electrocution, Article 680 on swimming pools has special grounding and bonding requirements to protect against stray currents, touch voltage, and step voltage.

Chapter 7 applies to special conditions such as emergency and standby power systems, fire alarm systems, optical fiber cables and raceways, and low-voltage systems. It modifies the general rules of Chapters 1–4 for these types of installations that have different characteristics than typical electric power systems. Example: Because the risk of shock is reduced at lower voltages, Article 725 doesn't require outlets, splices, and junctions of Class 2 and Class 3 circuits to be in boxes, which 300.15 (in Chapter 3) requires for wiring methods in general.

Chapter 8 applies to communications systems. It's a self-contained chapter that more or less "stands alone" outside the main body of *Code* requirements. The safety requirements of Chapters 1–7 do not apply to Chapter 8 systems unless they are specifically referenced in Chapter 8. This is discussed in greater detail in Unit 5: Other Systems.

Chapter 9 consists of tables giving reference information about conductors, raceways, and low-voltage power supplies. Chapter 9 applies as referenced elsewhere in the *Code*.

Annexes A–G contain reference information such as calculations, examples, and tables but aren't enforceable parts of the *NEC*; a statement to this effect appears at the beginning of each annex. The three most commonly used annexes are:

- *Annex C—Conduit and Fill Tables for Conductors and Fixture Wires of the Same Size*: Multiple conductors of the same size grouped together in a raceway is a common circuit configuration.

- *Annex D—Examples*: This contains examples of load calculations for services, feeders, and branch circuits.
- *Annex H—Administration and Enforcement* contains rules for adopting and enforcing the *National Electrical Code*. It is intended for use by states, cities, and counties and doesn't contain information needed by engineers and designers.

NEC Numbering System

The basic numbering system of the *Code* is as follows. More detailed information about the parallel numbering system for cable and raceway articles is included in Unit 3.

- Article 90—Introduction
- Chapter 1—General
  - Articles 100–110
- Chapter 2—Wiring and Protection
  - Articles 200–285
- Chapter 3—Wiring Methods and Materials
  - Articles 300–398
- Chapter 4—Equipment for General Use
  - Articles 400–490
- Chapter 5—Special Occupancies
  - Articles 500–590
- Chapter 6—Special Equipment
  - Articles 600–695
- Chapter 7—Special Conditions
  - Articles 700–770
- Chapter 8—Communications Systems
  - Articles 800–830
- Chapter 9—Tables
  - No articles

Subdivisions of the *NEC*

The *National Electrical Code* is organized into chapters, articles, parts, sections, and annexes. Except when discussing *Code* organization in this way, most users don't refer to "chapters" or "parts" very often. Most users think of the *NEC* in terms of articles and sections.

Chapters

Chapters are major subdivisions of the *NEC* that cover broad subject areas.

Articles

Chapters 1–8 are divided into articles, each of which covers a specific major subject, such as:

- Article 250—Grounding and Bonding
- Article 344—Rigid Metal Conduit: Type RMC
- Article 408—Switchboards and Panelboards
- Article 515—Bulk Storage Plants
- Article 760—Fire Alarm Systems

Chapter 9 consists of tables, and is not divided into articles.

Scope

Each article has a scope, which is always the first numbered section:

354.1 Scope
This article covers the use, installation, and construction specifications for nonmetallic underground conduit with conductors (NUCC).

Parts

Many articles are divided into parts that correspond to logical groupings of information. Parts have titles and are designated by roman numerals, for example, *I. Installation, II. Construction Specifications,* and *III. Grounding.* Parts typically consist of multiple sections.

Sections

Each section of the *NEC* can be thought of as representing a separate rule. Sections have boldface titles and are designated using the article number and a period, plus the individual section number:

- 110.23 Current Transformers
- 230.23 Size and Rating
- 348.56 Splices and Taps
- 450.5 Grounding Autotransformers
- 680.52 Junction Boxes and Other Enclosures

Note

It is becoming less common to use the word "section" when referring to numbered *NEC* rules. This book follows modern practice by generally referring to rules by number alone [210.8(A)] or as follows: 695.4. It uses the word "section" only at the beginning of a sentence.

Smaller Subdivisions

Some sections are in turn divided into smaller units, such as 230.23(C), 450.5(A)(3), and 680.52(B)(1)(a). However, generally speaking, all numbered *Code* subdivisions below the level of part are referred to as "sections." Terms such as "subsection" or "paragraph" aren't commonly used, and they don't appear in the *NEC Style Manual*, the official guide for writing the *National Electrical Code*.

Code Language

The *NEC* is intended for regulatory enforcement by states and local jurisdictions. It consists of mandatory rules, permissive rules, and explanatory information [90.5]. These three classes of information are presented in various ways.

Mandatory Rules

Mandatory rules describe actions that are required or prohibited. They use the terms *shall, shall not,* and *shall not be.*

Permissive Rules

Permissive rules describe actions that are permitted but not required, such as options or alternate methods. They use the terms *shall be permitted* and *it shall be permissible.* The *NEC* doesn't use the words "may" or "can."

Fine Print Notes

A Fine Print Note (FPN) contains explanatory information, such as references to other standards or information related to a *Code* rule. FPNs are not requirements, and aren't enforceable parts of the *National Electrical Code*.

Fine Print Note to 110.12
FPN: Accepted industry practices are described in ANSI/NECA/I-2006, *Standard Practices for Good Workmanship in Electrical Contracting,* and other ANSI-approved installation standards.

Exceptions

Exceptions are either mandatory rules or permissive rules, depending on the type of language used.

Tables and Figures

Tables and figures in the *Code* generally represent mandatory requirements (**Figure 1.3**). Tables and figures located in annexes, or identified as FPNs, are intended only to illustrate typical situations, and do not represent mandatory requirements.

NEC Is a Permissive Document

As discussed at greater length in Unit 3, the *National Electrical Code* is a safety standard and not a design guide. Rather than attempt to specify every detail in every possible type of installation, the *Code* provides a large number of general rules. Engineers, designers, and

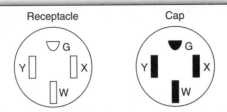

Receptacle   Cap

125/250-V, 50-A, 3-pole, 4-wire, grounding type

The attachment plug cap shall be a 3-pole, 4-wire, grounding type, rated 50 amperes, 125/250 volts with a configuration intended for use with the 50-ampere, 125/250-volt receptacle configuration. It shall be listed, by itself or as part of a power-supply cord assembly, for the purpose and shall be molded to or installed on the flexible cord so that it is secured tightly to the cord at the point where the cord enters the attachment plug cap. If a right-angle cap is used, the configuration shall be oriented so that the grounding member is farthest from the cord.

Figure 1.3 Attachment plug caps for mobile and manufactured homes. Reprinted with permission from 2005 NEC Handbook, copyright © 2005, National Fire Protection Association.

installers then apply these general wiring rules to a large number of specific real-world situations.

For this reason, the *NEC* can be thought of as a "permissive" wiring code. If a practice isn't specifically prohibited by the *Code*, it is permitted.

No Mandatory References to Other Industry Standards

The *NEC* is a self-contained safety code. It contains no mandatory references to other industry codes or standards. All references to other standards are in FPNs and are informational in nature. The relationship between the *National Electrical Code* and other industry codes and standards is discussed further in Unit 6 of this book.

Conclusion

This unit discusses the basic structure, organization, and language of the *National Electrical Code*, as well as its place in the U.S. standards system. Content, including regulatory and technical concepts that influence how engineers use NFPA 70 as one resource in designing electrical and communications systems for buildings and similar structures, are discussed in subsequent units of this book.

UNIT 2

General Considerations

Introduction

The previous unit described structure and organization of the *National Electrical Code,* along with an explanation of how the *Code* uses language. This unit extends that discussion to begin talking about general concepts engineers must understand to use the *NEC* as one tool in the process of designing electrical systems for buildings and similar structures. Much of it deals with how NFPA 70, a voluntary industry standard written by a fire-protection organization, is adopted and enforced by state and local governments as their official wiring rules. This unit covers the following:

- *NEC* Purpose
- *NEC* Scope
- Regulatory Adoption
- Approval and Enforcement
- Conditions of Installation and Use
- Mechanical Execution of Work (Workmanship)
- Product Listing
- UL *White Book*
- Industrial Exemptions
- Definitions
- Units of Measurement
- How the *NEC* Is Related to International Standards

NEC Purpose [90.1]

The *National Electrical Code* isn't a performance standard for designing electrical installations. Instead, it's a collection of safety rules intended to protect people and property from electrical hazards.

90.1 Purpose
(A) Practical Safeguarding. The purpose of this *Code* is the practical safeguarding of persons and property from hazards arising from the use of electricity.

The *Code* contains minimum provisions considered necessary for safety, which generally fall into three areas:

1. Preventing electrical shock
2. Preventing fires of electrical origin
3. Preventing injuries and property damage due to falling electrical equipment

Not a Design Guide

The *NEC* is neither a design specification nor an instruction manual. Every rule in it has a safety justification, and proposed revisions that lack a strong safety rationale are nearly always rejected. This makes the *Code* a conservative document that changes slowly, which is appropriate for a regulatory code adopted by over 42,000 state, county, and municipal jurisdictions across the United States.

The *Code* doesn't contain pure design requirements whose only purpose is convenience, aesthetics, energy efficiency, quality, or adequacy of the electrical installation for future expansion and changes. Both the FPN 90.1(B) and 90.8 address this subject (**Figure 2.1**).

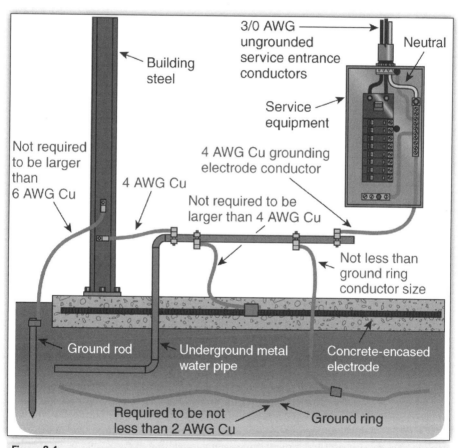

Figure 2.1 Code rules deal with the safety-related performance of electrical premises wiring systems. Reprinted with permission from 2005 NEC Handbook, copyright © 2005, National Fire Protection Association.

Safety Minimums

Thus, engineers designing electrical installations for buildings and similar structures should regard *NEC* rules as the minimum requirements for safety, but not state-of-the-art design guidance. An electrical installation built to the *Code* will be safe—but it may be incomplete in certain respects, may not perform adequately, and may not meet the needs of its end-users.

This is because the *NEC* is a safety code, rather than a design manual. Electrical systems designs must always start with *Code* requirements, but typically don't end there. Many electrical installations exceed *NEC* minimums in many respects, largely because these don't include many practical details of typical electrical installations. Example: There's no *NEC* rule requiring that light switches be located near doorjambs, on the side opposite the hinges, although this design practice is nearly universal.

Electrical engineers and designers must comply with the minimum safety rules of the *NEC*, but they are always free to exceed these minimums. Nothing in the *Code* restricts the use of design or installation techniques that go beyond these minimums.

In addition, there are mandatory electrical requirements in other industry standards that apply to the design and installation of electrical systems in buildings and similar structures. This is covered in greater detail in Unit 4 of this book.

NEC Scope

The *Code* covers installation of electrical and communications systems in buildings and similar structures, along with associated outdoor areas such as parking lots [90.2(A)]. It doesn't cover the following [90.2(B)]:

- Vehicles, aircraft, or boats and ships (but the *NEC* does cover floating buildings such as restaurants and casinos, in Article 555)
- Installations in surface or underground mines
- Electric utility generating, transmission, and distribution systems

Voltage Levels

Most *Code* rules cover electrical systems operating at 600 volts or less. There are also a limited number of rules for installations operating above 600 volts. These high-voltage rules of the *NEC* apply to industrial installations, campus-style electrical distribution systems, and in special circumstances such as neon-tube and cold-cathode lighting installations. This book applies to electrical equipment and systems operating at 600 volts or less, unless otherwise noted.

Electric utility generating, transmission, and distribution systems aren't within the scope of the *NEC*. Instead, they are covered by a different document with a confusingly similar name, the *National Electrical Safety Code* (*NESC*), ANSI/IEEE standard C2-2007.

Differences Between the *NEC* and *NESC*

The rules of the two electrical codes for systems operating above 600 volts are different, in many respects. In general, the *NESC* permits smaller-sized conductors and raceways than the *NEC* to serve the same loads, and has fewer rules for overcurrent protection and disconnecting means.

The primary explanation for these differences is that electric utilities have employees to monitor and maintain their systems, and thus are able to respond quickly when operational problems, such as short circuits, ground faults, fallen utility poles, or storm damage, occur. Many owners of *NEC*-based electrical systems don't have this capability. For this reason, there is more "safety factor" built into *NEC* wiring rules, as compared to the *NESC*.

Service point. The dividing point between electric utility distribution lines and customer-owned premises wiring is called the *service point*. This term is defined identically in both the *NEC* (Article 100—Definitions) and the *NESC* (Section 2: Definitions). Typically, the *service point* is at the utility watt-hour meter (**Figure 2.2**).

- Conductors and equipment on the line side of the meter (upstream) are owned by the utility and fall under the jurisdiction of the *NESC*.

- Conductors and equipment on the load side of the meter (downstream) are owned by the customer and fall under the jurisdiction of the *NEC*.

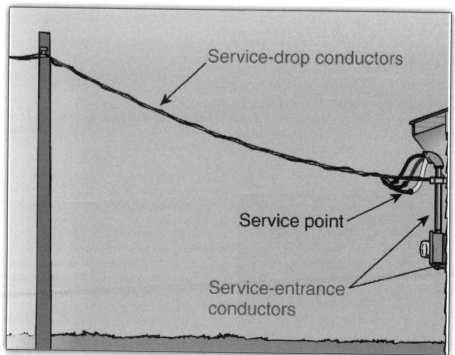

Figure 2.2 "Service point" is a key concept in both the *NEC* and the *National Electrical Safety Code*. It is normally located at or near the utility watt-hour meter. Reprinted with permission from 2005 NEC Handbook, copyright © 2005, National Fire Protection Association.

Street and area lighting. Outdoor lighting for roadways and parking lots is a "gray area" that sometimes falls under the *NEC* and sometimes under the *NESC*. Roadway lighting on public property or utility rights-of-way generally isn't considered to be within the scope of the *NEC* [90.2(B)(5)(b)], and typically is installed under *NESC* rules.

Section 90.2(A) states that yards, lots, and parking lots are within the scope of the *National Electrical Code,* thus implying that all outdoor lighting located on private property must be installed to *NEC* rules. However, some local jurisdictions permit electric utilities to install and service outdoor lighting on private property.

Regulatory Adoption

The *National Electrical Code* is adopted for regulatory use at both the state and local level, but the way in which this happens isn't uniform across the country.

- Some states adopt the *Code* for all electrical construction within their limits.
- Some states don't adopt the *NEC,* and allow (but don't require) cities and counties to do so.
- Some states adopt the *Code,* but allow cities and counties to enforce more (but not less) stringent electrical rules at their local option.

Annex H—Administration and Enforcement contains procedures for adopting and enforcing the *National Electrical Code.* (Its clauses are numbered 80.1 through 80.35 because Annex G was previously Article 80 of the *NEC.*)

Effective Date

Although many jurisdictions adopt each new edition of the *National Electrical Code* as soon as it is officially available (for example, adopting NFPA 70-2008 on January 1, 2008), this practice isn't universal. Many state and local governments take longer to adopt the latest *NEC,* due to factors such as slow regulatory processes and the need to train public employees, such as electricians and electrical inspectors, in the new *Code* rules.

Note

Each new edition of the *National Electrical Code* is actually published several months before its cover date (usually in September) to allow users, inspectors, trainers, and others to become familiar with the changed requirements. However, states and local jurisdictions don't adopt the *Code* for regulatory enforcement prior to its effective date.

This means many states, cities, and counties routinely operate by enforcing the requirements of older versions of the *NEC.* Many jurisdictions are one, two, or even more editions behind the current NFPA 70-2008. Engineers designing electrical systems for buildings and similar structures should check with their state or local building department to verify which edition of the *NEC* is in force. The National Electrical Contractors Association (NECA) also maintains a list at www.neca-neis.org/state.

Local Variances

Although many jurisdictions adopt the *National Electrical Code* as published in its entirety, others adopt it with local changes. A fairly common one is requiring minimum 12 AWG conductor sizes for commercial wiring, while allowing 14 AWG conductors only in one- and two-family dwellings. (This also has the effect of requiring minimum 20-ampere branch circuits for commercial applications, while allowing 15-ampere branch circuits only in one- and two-family dwellings.)

The *NEC* allows minimum 14 AWG conductors and 15-ampere branch circuits for all general wiring applications, not restricting them to residential occupancies. Engineers should check with their state or local building department to verify whether the *NEC* was adopted *in toto* or with local variances.

Approval and Enforcement

Engineers, designers, and others often talk about electrical products or applications as being "*NEC*-approved," "*Code*-approved," or "UL-approved," but all of these expressions are inaccurate. The authority having jurisdiction (AHJ) is responsible for approving installations, approving electrical products and materials, interpreting *NEC* rules, granting the special permission mentioned in a number of rules, and in some cases waiving specific *Code* requirements [90.4].

Product listing is an important component that provides a basis for AHJ approval. It is discussed later in this unit.

What Authority Has Jurisdiction?

The authority having jurisdiction is the entity responsible for enforcing regulatory codes, conducting inspections, reviewing plans (in some jurisdictions), and approving electrical installations. Having said this, exactly *who* the AHJ is differs widely, depending on the location and situations.

Departmental names. Responsibility for regulating electrical construction falls under different government agencies in different places, as shown in **Table 2.1**. This book uses the term *building department*.

Building officials, electrical inspectors, combination inspectors. *NEC* Article 100 defines *authority having jurisdiction* as "The organization, office, or individual responsible for approving equipment, materials, an installation, or a procedure." *Building official* is a generic term (not defined in the *Code*) that indicates the person in charge of this organization or office, usually a nonelectrical administrator. Although some building officials are engineers, architects, or former inspectors, others aren't.

Table 2.1 Authorities Having Jurisdiction over Electrical Construction

Alaska	Department of Community and Economic Development
Connecticut	Department of Consumer Protection
Iowa	Department of Public Safety
Minnesota	State Board of Electricity
Nebraska	State Electrical Division
New Jersey	Department of Community Affairs
New Mexico	Construction Industries Division
Vermont	Department of Labor & Inspection
Wyoming	Department of Fire Prevention and Electrical Safety

- *Electrical inspectors* (another term not defined in the *NEC*) typically work for and report to building officials.
- *Combination inspectors.* States and local jurisdictions are increasingly using combination, or "three-hat" inspectors, to inspect residential projects and even smaller commercial construction projects such as strip shopping centers. Rather than have separate structural, mechanical/plumbing, and electrical inspectors, one cross-trained individual inspects the entire building.

State or local? Some states have centralized building departments for all construction within the state; North Carolina and Connecticut are examples. Other states have no building departments, and leave this regulatory function to local jurisdictions: cities, towns, counties, villages, and townships. Georgia and Louisiana take this approach. Many states use a mixed system. Some local jurisdictions, typically the larger and more urban ones, have their own building departments, while the statewide building department has responsibility for inspecting and approving electrical construction in unincorporated rural areas. Michigan and Wisconsin both follow this model.

However, the reality is that *Code* enforcement and inspection are nearly nonexistent in many rural areas. For example, only 5 of Alabama's 67 counties regulate electrical construction, although cities such as Birmingham and Montgomery have their own building departments. A great deal of electrical work in rural areas is performed by unlicensed and often untrained people characterizing themselves as electricians and contractors, but whose competence has not been tested by any government entity.

Third-party inspection agencies. Some jurisdictions don't maintain their own building departments for reasons of cost or low construction activity. Instead, they contract out their inspection services to private agencies. For example, Middle Department Inspection Agency (MDIA) is a private company based near Philadelphia that provides building department services to more than 200 cities, towns, and counties throughout Pennsylvania, New York, Delaware, Maryland, Virginia, and West Virginia.

Table 2.2 Emergency Systems	
Fire alarm systems, mass notification systems	NFPA 72, *Standard for Installing Fire Alarm Systems*
Emergency egress pathways and lighting, exit signs	NFPA 101, *Life Safety Code*

Fire departments. Fire department fire marshals' offices and local fire departments normally have inspection and approval authority over fire alarm systems, emergency egress pathways and lighting, exit signs, and mass notification systems (where required). These items are not under the jurisdiction of the *NEC*, but under the other standards shown in **Table 2.2**. The relationship between the *National Electrical Code* and other building codes and standards is discussed at great length in Unit 5.

Industrial occupancies. Large industrial plants are often exempt from state, city, or county inspection authorities, because their equipment and systems are so different from those installed in normal commercial and industrial buildings such as office towers, shopping malls, and schools. In these cases, the AHJ may be an employee of the industrial plant (e.g., the plant engineer or designee) or a third-party inspection agency contracted to provide this service.

Major companies. When large national companies build facilities in areas without state or local building departments, they typically require that all electrical construction be performed in accordance with the current edition of the *National Electrical Code*. This is written into their corporate plans, specifications, and construction guidelines. In such cases, the AHJ may be an employee of the customer's construction department or a third-party inspection agency.

Electric utilities. In rural areas without building departments, the serving electric utility frequently conducts a service inspection to verify that the service equipment (switchboard or main distribution panelboard) has been installed correctly and safely. However, the utility does not later inspect the premises wiring to ensure that it has been installed in accordance with the *National Electrical Code*.

Permits, Plan Review, and Inspections

Nearly all urban and suburban jurisdictions require building permits for electrical construction work. The permit fees finance inspections by the AHJ. Building permits of the type shown in **Figure 2.3** contain a general description of the work authorized to be performed (for example, electrical, plumbing, structural) and must be posted at the job site. Typically, there are at least two inspections during an electrical construction job: rough and final (or finish). Often, plan review precedes these field inspections.

Plan review. Prior to issuing a permit, many building departments require that electrical plans and load calculations be submitted; they are reviewed to ensure that the design complies with *NEC* requirements for that type of occupancy. Smaller jurisdictions often don't

BUILDING PERMIT

Any Town, Any County
Department of Inspection Services
987-654-3210

The Department of Inspection Services hereby grants permission to

_____

_____

_____

_____

_____

Approved by _____ , Director of Inspection Services   Date: _____

PERMIT NO.: _____   By _____ , Permit Clerk

This permit conveys no right to occupy any street, alley or sidewalk, or any part thereof, either temporarily or permanently except that specifically provided for in the building code. It is the owner's responsibility to determine that the proposed construction does not violate existing private covenants and/or subdivision restrictions applicable to subject property or violate any State, Municipal or County zoning, subdivision or health department law, rule, ordinance, or regulations.

Approved plans MUST be retained on job and this card KEPT POSTED until final inspection has been made. Such building SHALL NOT BE OCCUPIED until FINAL INSPECTION has been made and approved.

POST THIS CARD

BUILDING INSPECTION APPROVALS	PLUMBING INSPECTION APPROVALS	ELECTRICAL INSPECTION APPROVALS
FOOTING OR FOUNDATIONS DATE _____ INSPECTOR _____ SLAB DATE _____ INSPECTOR _____ FRAMING–PRIOR TO SHEETROCK OR LATH DATE _____ INSPECTOR _____ FINAL INSPECTION DATE _____ INSPECTOR _____	1. PLUMBING IN SLAB 　DATE _____ INSP _____ 2. ROUGH DRAINAGE AND VENTS 　DATE _____ INSP. _____ 3. ROUGH WATER IN BUILDING 　DATE _____ INSP. _____ 4. BUILDING SEWER 　DATE _____ INSP. _____ 5. FINAL INSPECTION 　DATE _____ INSP. _____	1. ELECTRICAL TEMPORARY 　DATE _____ INSP. _____ 2. ROUGH ELECTRICAL 　DATE _____ INSP. _____ 3. FINAL INSPECTION 　DATE _____ INSP. _____ **GAS INSPECTION APPROVALS** 1. ROUGH GAS PIPING 　DATE _____ INSP. _____ 2. FINAL GAS PIPING 　DATE _____ INSP. _____ **MECHANICAL INSPECTION APPROVALS** 1. GAS VENT 　DATE _____ INSP. _____ 2. FINAL MECHANICAL 　DATE _____ INSP. _____
MARKS	**WORK SHALL NOT PROCEED UNTIL EACH DIVISION HAS APPROVED THE VARIOUS STAGES OF CONSTRUCTION.**	
EXCAVATORS SHOULD CALL LINE LOCATION CENTER 1-800-666-3000 48 HOURS BEFORE DIGGING	THE INSPECTION BY THE CITY OR COUNTY OF THIS BUILDING CONSTITUTES NO REPRESENTATION EXPRESS OR IMPLIED BY THE CITY OR COUNTY OR ITS EMPLOYEES AS TO THE COMPLIANCE OF THIS BUILDING WITH THE REQUIRED SET-BACK LINES OR THE QUALITY OF THE WORKMANSHIP OR MATERIALS USED THEREIN.	

UNLAWFUL TO REMOVE OR DEFACE THIS CARD
UNTIL CONSTRUCTION IS COMPLETE

Figure 2.3 A building permit must be posted at the job site during the period of construction.

conduct plan reviews before issuing permits, and many jurisdictions don't require them for one- and two-family dwellings.

Rough inspection. The first inspection takes place when the service has been installed, outlets (boxes) are in place, and wiring is complete—but before the walls have been "closed in" and receptacles, switches, luminaires and other electrical equipment have been installed. This allows the inspector to verify that the service, branch circuits, and outlet locations comply with *National Electrical Code* requirements. On larger and more complex projects, there may be several rough inspections at different stages of the job, such as before concrete is poured over raceways in slabs.

Finish (final) inspection. The second (or last) inspection takes place when the electrical installation is complete and power has been turned on. This allows the inspector to verify that the electrical system operates safely, that receptacles are wired properly, and that all other applicable *Code* requirements have been met.

What is "red tagging"? When inspectors find *Code* violations on a job in progress, they typically issue a stop-work order with list of deficiencies to be corrected. Often this form is posted at the service, or next to the electrical permit, and is typically called a *red tag* in the field (though it may not actually be red). Normally, the local jurisdiction will grant an occupancy permit—and the local utility will turn on power—only when all deficiencies have been corrected and the AHJ has withdrawn the stop-work order and reinstated the building permit.

Electrician and Contractor Licensing

Many jurisdictions require that only licensed contractors and electricians perform work within the scope of the *NEC*. Some that require licensing provide an exception that allows owners and/or their employees (such as building engineers and maintenance personnel) to perform electrical work on the owner's own property. Engineers who perform maintenance, testing, and troubleshooting work on electrical products and systems must understand the applicable licensing laws in the areas where they work.

Qualified person. The *National Electrical Code* doesn't require that work be performed by licensed electricians or contractors. This is because, in general, the *Code* describes *what* must be done to ensure electrical safety, but not *who* must do the work. Some types of work covered by *NEC* rules are typically performed by engineers, HVAC technicians, or other specialized installers and maintenance personnel who, although not electricians, are technically competent to perform the work safely. Lastly, not all states and municipalities require licensing of electrical installers.

Rather than addressing licensing or training, the *Code* often refers to work being done by a qualified person. Article 100 defines *qualified person* as "one who has skills and knowledge related to the construction and operation of the electrical equipment and installations and has received safety training on the hazards involved."

Conditions of Installation and Use

NEC 110.2 states that conductors and equipment are acceptable only if approved, and Article 100 defines *approved* as "[a]cceptable to the authority having jurisdiction." To further explain factors that go into AHJ approval of electrical installations, the FPN lists several related *Code* sections:

- 90.7 Examination of Equipment for Safety
- 110.3 Examination, Identification, Installation, and Use of Equipment
- Definitions of *Approved, Identified, Labeled,* and *Listed* [Article 100]

NEC 110.3(B) states that listed or labeled equipment must be installed and used in accordance with any instructions included in the listing or labeling [110.3(B)]. This is actually one of the most important rules in the *NEC* because it means that product listing and labeling requirements are enforceable rules of the *National Electrical Code*. This is discussed further in the section on product listing later in this unit.

Hospital Grade Receptacles

As an example, the following question-and-answer discussion illustrates how 110.3(B) works in practice:

Q: What *Code* rule requires "hospital-grade" receptacles with isolated grounds to be marked with a green dot?

A: The answer to this question actually has two parts.

Part 1: Green dot receptacles. Hospital-grade receptacles are required by 517.18(B) for all receptacles provided at patient-bed locations. Each hospital-grade receptacle must be grounded by means of an insulated copper conductor.

However, the *NEC* rule doesn't mention green dots; instead, that requirement is part of the product listing requirement. UL lists hospital-grade receptacles in a category called Receptacles for Plus and Attachment Plugs (RTRT). Guide information in the UL *White Book* states that receptacles for hospital use in accordance with *NEC* Article 517 are identified by the marking "hospital grade" and a green dot on the face of the receptacle.

Thus, all hospital-grade receptacles that comply with 517.18(B) must have a green dot, although this requirement doesn't appear in actual *Code* wording. This is a good illustration of 110.3(B), which in effect makes product listing requirements a part of enforceable *National Electrical Code* rules:

 110.3(B) Installation and Use Listed or labeled equipment shall be installed and used in accordance with any instructions included in the listing or labeling.

Part 2: Isolated ground receptacles. Section 517.18(B) requires that each hospital-grade receptacle must be grounded by means of an insulated copper conductor. The intent of this rule is to ensure that these receptacles have a reliable grounding path back to the supplying

panelboard. Fluids and flammable gases are often used at patient-bed locations, increasing the potential risk of electric shock.

However, this isn't the same thing as an "isolated ground receptacle," which is permitted for the reduction of electrical noise (electromagnetic interference) by 250.146(D) and 406.2(D). The UL *White Book* (Category RTRT) defines it as a grounding-type receptacle with grounding terminals purposely insulated from the mounting means and metal cover plates. Section 406.2(D) requires isolated ground receptacles to be identified by an orange triangle on the face of the receptacle. (Normally, they are manufactured from orange plastic, with a triangle molded into the receptacle face.)

Mechanical Execution of Work (Workmanship)

Section 110.12 requires that electrical equipment be installed in a "neat and workmanlike manner." This is one of the most argued-about rules in the *National Electrical Code*. It establishes a performance requirement without providing sufficient additional details for clear enforcement. For this reason, 110.12 is the basis of many disputes between design engineers, installers, customers, and AHJs.

Other articles throughout the *Code* also contain sections on workmanship that provide additional installation details for particular wiring methods or types of equipment.

- 110.12—General
- 640.6 and 640.22—Audio Systems
- 725.8—Class 1, 2, and 3 Circuits
- 760.8—Fire Alarm Systems
- 770.24—Fiber Optics
- 800.24—Communication Circuits
- 820.24—CATV (Coaxial Cable)
- 830.24—Broadband Systems

Fine print notes. Most of these *Code* sections are followed by a FPN that lists other ANSI-approved installation standards as sources of information on "accepted industry practices." Although these FPNs aren't enforceable provisions of the *NEC*, engineers designing premises wiring systems should be familiar with these references.

Product Listing

Regulatory enforcement of *National Electrical Code* rules depends on the listing of products by testing laboratories. Independent, third-party evaluation of electrical equipment [90.7, 110.3] is an essential element of what is often called the "U.S. electrical safety system."

Most states, cities, and counties don't have testing laboratories or personnel trained to evaluate equipment, but product listing provides a uniform basis for electrical inspectors in the field to approve products and materials as complying with the *NEC*.

Listing means essentially the same thing as *certification*, but listing is the term commonly used in the United States, because certification carries with it a sense of approval. As discussed, only the AHJ can approve products, materials, and installations under the rules of the *National Electrical Code*. Listing agencies don't approve products or materials; they merely test and list them.

Product listing can typically be summarized as follows:

- Safety requirements that affect the design of an electrical product are written into the *Code*.
- Listing agency utilizes a *product safety standard*, typically an ANSI standard, governing the safety aspects of how that product shall be constructed.
- Listing agency tests representative samples of the product to verify compliance with the standard.
- Products manufactured in accordance with the standard are marked by the manufacturer with the listing agency's label or "mark."
- Listing agency publishes a directory of listed products. This directory provides more information than can be placed on the actual label or mark, which is typically small.
- Electrical inspectors seeing labeled products in the field consult the directory to verify that the products conform to *NEC* and product use requirements for a particular application.
- The listing agency performs follow-up inspections and testing at intervals, on actual listed commercial products, to verify continued compliance with the product safety standard.
- Titles of ANSI-approved product safety standards are listed in Annex A–Product Safety Standards.

Underwriters Laboratories Inc.

There are a number of different listing agencies for different types of construction products and materials. The leading electrical listing agency is Underwriters Laboratories Inc. (UL), because it is also an ANSI-accredited standards developer that publishes more than 1000 safety standards for electrical construction products of all kinds, ranging from circuit breakers and panelboards to raceways and cables to luminaires.

There are also other electrical listing agencies, some of which concentrate on particular niches such as data cabling, air conditioning equipment, or industrial products, but the others aren't accredited standards developers. Instead, they test products to ANSI/UL standards or a few other ANSI-approved standards. This can be seen in *NEC* Annex A, where over 90 percent of the 200-plus product safety standards are published by UL.

For this reason, the UL mark is the one most often seen on electrical construction products and utilization equipment. UL's electrical equipment listing directory, commonly known as the *White Book*, is an important resource for electrical engineers and designers. It is discussed later in this unit.

Nationally Recognized Testing Laboratories

Product listing organizations are sometimes referred to collectively as nationally recognized testing laboratories (NRTLs, pronounced *nertles*). However, this term doesn't appear in the *NEC*. OSHA approves NRTLs and refers to them in its Part 1910 and 1926 electrical regulations. The term also appears in NFPA 70E-2004, *Standard for Electrical Safety in the Workplace*.

The *NEC* uses the term "listed," which is defined in Article 100. It doesn't specify that the listing laboratory must be a NRTL, because that designation refers only to its certification by OSHA to conduct safety testing for products used in the workplace.

UL *White Book*

UL's *Guide Information for Electrical Equipment* (*White Book*) is an indispensable reference for electrical engineers and designers who select and specify electrical construction and utilization equipment (**Figure 2.4**). The exact information provided varies from one product to the next, but generally includes the following:

- **General.** Detailed product description information on use and application, electrical ratings, related *NEC* article(s), and/or other regulatory standards.

- **Requirements.** The UL standard(s) under which the products are listed. Common UL *White Book* language is "The basic standard used to investigate products in this category is UL 999."

- **UL Mark.** Describes what information appears on the product. Typically this includes the UL symbol together with the word LISTED, a control number, and a product name such as "Electric Sign" (**Figure 2.5**).

- **Listing Categories.** Each type of product has a four-letter category code; these are not acronyms related to product names. Examples:
  - Armored Cable (AWEZ)
  - Luminaires and Fittings (HYXT)
  - Fuses (JCQR)
  - Panelboards (QUEY)
  - Rigid Ferrous Metal Conduit (DYIX)
  - Rigid Nonferrous Metallic Conduit (DYWX)
  - Switches, Enclosed (WIAX)
  - Transformers, General Purpose (XPZZ)

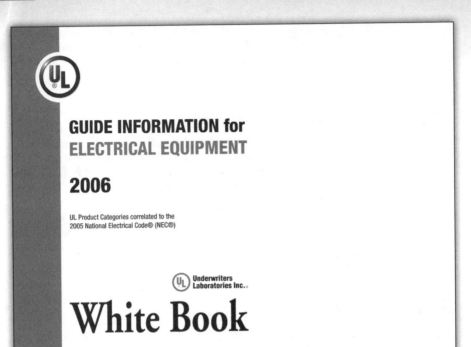

Figure 2.4 UL's *Guide Information for Electrical Equipment* (*White Book*) is a basic reference for electrical engineers and designers. (Reprinted with permission from Underwriters Laboratories Inc.)

UL Marks

UL has many different marks for different categories of products. The three most common ones are the Listing mark, Recognized Component mark, and Classification mark.

Listing mark. The familiar UL in a circle indicates that a product has been tested for safety and compliance with specific safety standards for installation in accordance with the *NEC* and/or other regulatory standards. The listing mark is applied to end-use products such as luminaires, panelboards, transformers, wiring devices, and appliances.

Recognized Component mark. The backwards UR appears on components such as relays, pushbuttons, terminal blocks, and thermostats that are intended to be factory assembled into other products (**Figure 2.6**). It isn't a listing mark and doesn't normally appear on end-use products.

> **Note**
> Products can be both listed and classified. Metallic outlet boxes are also listed for safety in category QCIT2 UL 514A "Metallic Outlet Boxes."

Classification mark. UL evaluates most products for safety only, testing them for characteristics such as fire and shock hazard. The listing mark indicates that products

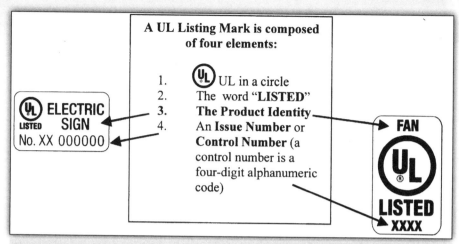

Figure 2.5 UL listing marks provide evidence that products have been evaluated for compliance with the *NEC.* (Reprinted with permission from Underwriters Laboratories Inc.)

Figure 2.6 The UL recognized component mark appears on components that are parts of end-use products. (Reprinted with permission from Underwriters Laboratories Inc.)

comply with the applicable safety standard and can be installed in accordance with the *NEC* and/or other regulatory safety standards, but these marks aren't related to product performance or quality. By contrast, the verification mark indicates that products comply with industry performance specifications rather than regulatory safety codes. It consists of includes the UL symbol together with the word VERIFIED, a control number, the product name, and the designation and date of the specification. Example: Optical Fiber Branching Devices (QBEN) are verified to Telcordia Specification GR-2866-CORE (Issue 1, June 1995).

NEC-UL Cross Reference

One useful feature of the *White Book* is the "Index of UL Product Categories Correlated to the 2005 NEC." It lists *Code* sections in numerical order along with the corresponding UL Product Category Code and *White Book* page number. Although not every numbered *NEC* section has corresponding listings, there are over 4,100 cross-reference entries in this index.

Marking Guides

The *White Book* includes seven marking guides for listed products. Although these are intended primarily to help AHJs and installers understand the meaning and location of markings on UL-listed products, they also contain technical and application information that can be useful to electrical engineers and designers. The seven marking guides are:

1. Molded Case Circuit Breakers
2. Dead-Front Switchboards
3. Electrical Heating and Cooling Equipment
4. Luminaires
5. Panelboards
6. Swimming Pool Equipment, Spas, Fountains and Hydromassage Bathtubs
7. Wire and Cable

These marking guides supplement the more detailed guide information in the UL *White Book*, but aren't a substitute for it. **Figure 2.7** shows the cover from the UL-marking guide for luminaires.

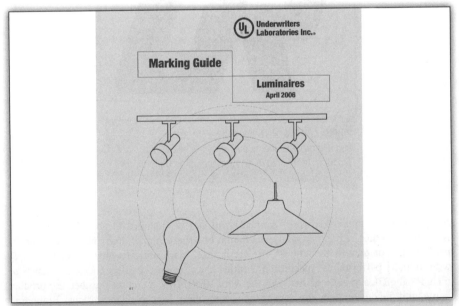

Figure 2.7 UL product marking guides contain useful technical and application information. (Reprinted with permission from Underwriters Laboratories Inc.)

Figure 2.8 The CE marking appears on products built to European standards.

Formats

UL's *White Book* is available as a paper directory, on CD, and can be downloaded at www.ul.com/regulators/2006WhiteBook.pdf or 2007 book at www.ul.com/regulators/2007WhiteBook.pdf. (Note: This URL is case sensitive.) The electronic CD version of the *White Book* uses the same software platform as the electronic *National Electrical Code* and *NEC Handbook*; it is user-friendly and easily searchable. There is also a UL online certifications directory at www.ul.com/database that can be used for verifying UL product listings.

What Is CE Marking?

The electrical product market has changed substantially in recent years. Globalization has resulted in an influx of products manufactured outside of North America, and U.S.-based listing agencies have responded by becoming global operations. They test and list products for the U.S. marketplace to ANSI standards with test laboratories located overseas, in the countries where the products are manufactured. This system has been effective at facilitating global trade.

However, electrical equipment built to European (CENELEC) or IEC standards, or custom-built to foreign manufacturers' specifications, is increasingly being imported to this country. This equipment frequently carries a "CE marking" (**Figure 2.8**), which is sometimes referred to as a *CE listing*.

Unlike the logos of UL and other U.S. listing agencies, the CE marking is not a safety certification mark. It is based on declarations by manufacturers (i.e., self-certification) that their products comply with applicable standards of the European directives, rather than third-party evaluation. The CE marking also doesn't indicate that a product has been designed to operate safely in an *NEC* environment.

Industrial Exemptions

Although the *National Electrical Code*, as a safety document, doesn't contain every detail needed to design and specify an electrical construction project, it is a prescriptive document that specifies many details needed to design and specify safe electrical installations. Engineers, electricians, and others follow *Code* rules and tables (modified in some cases by diversity and derating factors) to select circuit breaker and fuse ratings, conductor ampacities, raceway sizes, motor overload protectors, wiring methods in hazardous (classified) locations, and such.

Most *Code* rules apply in a relatively uniform manner across different applications and occupancies. Section 230.71 permits a maximum of six disconnects per incoming service in both apartment buildings and shopping malls, and a 250 kcmil Type THW aluminum conductor will safely carry 205 amperes whether installed in a school or a water treatment plant.

However, the *Code* contains approximately 40 special rules that apply only in industrial occupancies and/or under engineering supervision. The assumption is that in manufacturing and process control industries where the electrical systems are installed and/or maintained by expert onsite crews consisting of qualified persons (as defined in Article 100), normal *NEC* safety rules can be more flexible than in typical commercial-institutional occupancies where an equivalent level of engineering and maintenance personnel are not available.

These 40-plus rules are often described as *industrial exemptions* or *industrial exceptions*. Typical *Code* language is "Where the conditions of maintenance and supervision ensure that only qualified persons service the installation. . . ." A list follows:

- 110.26(A)(1)(c)
- 200.6 Ex. No. 1
- 200.9 Ex.
- 210.3 Ex.
- 210.6(E)
- 210.9 Ex. No. 2
- 215.11 Ex. No. 2
- 240.2(1)
- 240.21(B)(4)(1)
- 240.21(C)(3)
- Article 240, Part VIII
- 250.21(3)b
- 250.36(1)
- 250.52(A)(1) Ex.
- 250.119(B)
- 250.122(F)(2)(1)
- 250.186(1)
- 336.10(7)
- 368.17(B) Ex.
- 392.3(B)
- 392.6(J)
- 396.10(B)
- 398.30(C)
- 410.130(G)(1) Ex. 4
- 427.22(1)
- 430.28 Ex.
- 430.102(B)(2)(b)
- 440.14, Ex.1

- 450.3(A) Table Note 3
- 500.7(K)
- 501.10(A)(1)(d)
- 501.140(A)(2)
- 502.10(A)(1)(3)
- 505.8(I)
- 505.15(B)(1)(c)
- 505.17
- 518.3(B) Ex.
- 590.6(A) Ex.
- 620.5
- 647.3(2)
- 685.1(2)
- 725.61(D)(4)
- 727.4
- 770.133(A) Ex. No. 3 & 4

Supervised industrial installation. Section 240.2 defines an industrial establishment as having three characteristics: a combined process and manufacturing load greater than 2500 kVA; at least one service rated 480Y/277 volts or greater; and conditions of maintenance and supervision ensuring that only qualified persons monitor and service the electrical system. All process and manufacturing loads (from low-, medium-, and high-voltage systems) can be added together to satisfy the 2500 kVA requirement, but loads not associated with manufacturing or processing cannot be included.

Strictly speaking, the definition in 240.2 applies only to Article 240, but it is generally used as the definition of a supervised industrial installation throughout the *National Electrical Code.*

Definitions

Article 100 contains definitions needed to understand and use the *NEC.* It doesn't include common electrical or technical terms such as *ampere, volt, generator, ohm, motor, circuit breaker, lamp, conduit,* and so forth.

Article 100 contains only definitions of terms that appear in at least two other *Code* articles. More specialized terms used in a single article are defined in that same article, usually in the second numbered section. For example, 517.2 contains definitions that apply only to health care facilities and 695.2 has definitions of specialized terms related to fire pump installations. Not every *NEC* article has its own definitions section.

How Definitions Relate to *Code* Rules

Definitions aren't *Code* rules *per se*; in fact, the *NEC Style Manual* prohibits expressing requirements in definitions. However, the way terms defined in Article 100 are used in mandatory *Code* language is limited to the sense of those definitions. So, although definitions aren't rules, understanding the Article 100 definitions is a critical first step to understanding and applying the *National Electrical Code*. Key definitions excerpted from Article 100 that engineers and designers must understand to properly apply the *NEC* are as follows. In some cases, they are also terms about which confusion exists in the field:

 Accessible (as applied to equipment). Admitting close approach; not guarded by locked doors, elevation, or other effective means.

Included here are equipment located behind an access door or panel that is not locked, luminaires installed in a dropped ceiling with lift-out panels, equipment that can be reached using a ladder, and busways with hookstick-operated switches.

 Accessible (as applied to wiring methods). Capable of being removed or exposed without damaging the building structure or finish, or not permanently closed in by the structure or finish of the building.

Examples of accessible wiring methods include junction boxes, cables, or busways located above a suspended ceiling with lift-out panels (**Figure 2.9**).

 Accessible, Readily (Readily Accessible). Capable of being reached quickly for operation, renewal, or inspections without requiring those to whom ready access is requisite to climb over or remove obstacles or to resort to portable ladders, and so forth.

These equipment and wiring methods are easier to reach than accessible ones. Readily accessible equipment and wiring methods can be located behind locked doors provided that persons who need ready access (such as maintenance electricians) have a key or lock combination. However, it must not be necessary to use a ladder or lifting device or remove an access panel to get at readily accessible equipment and wiring methods.

 Ampacity. The current, in amperes, that a conductor can carry continuously under the conditions of use without exceeding its temperature rating.

Strictly speaking, ampacity only applies to conductors. Equipment has ampere ratings. However, in practice, *Code* users frequently talk about the ampacity of circuit breakers, fuses, and the like. Many factors affect the ampacity ratings of conductors in the *NEC*, such as the temperature correction factors for ampacity in Tables 310.16 through 310.20 and Annex B.

 Approved. Acceptable to the authority having jurisdiction.

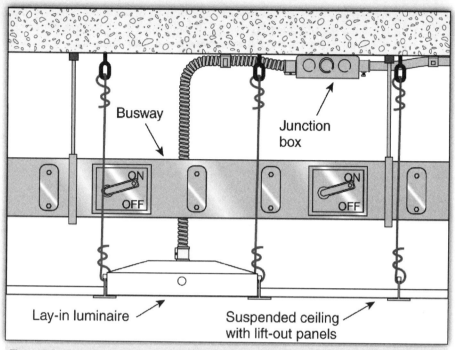

Figure 2.9 Wiring methods located behind removable panels or above suspended ceilings with lift-out panels are considered accessible as defined in Article 100.

This is an important definition for correctly applying the *NEC*. People often talk about a wiring method being "approved by the *Code*" or electrical equipment being "UL approved," but both expressions are inaccurate. Only the AHJ can approve wiring methods and equipment installed under the rules of the *NEC*; see the following definition. Also see the definitions of *labeled* and *listed*.

Authority Having Jurisdiction (AHJ). An organization, office, or individual responsible for enforcing the requirements of a code or standard, or for approving equipment, materials, an installation, or a procedure.

FPN: The phrase "authority having jurisdiction" is used in NFPA documents in a broad manner, since jurisdictions and approval agencies vary, as do their responsibilities. Where public safety is primary, the AHJ may be a federal, state, local, or other regional department or individual such as a fire chief; fire marshal; chief of a fire prevention bureau; labor department, or health department; building official; electrical inspector; or others having statutory authority. For insurance purposes, an insurance inspection department, rating bureau, or other insurance company representative may be the authority having jurisdiction. In many circumstances, the property owner or his or her designated agent assumes the role of the authority having jurisdiction; at government installations, the commanding office or departmental official may be the authority having jurisdiction.

Most often the AHJ is a government-employed electrical, inspector, however, as the long FPN to this definition explains, the AHJ may also be someone else, depending on the particular circumstances of an installation. AHJ is a widely used acronym that can mean either electrical inspector or building official.

Bonded (Bonding). Connected to establish electrical continuity and conductivity.

Some *Code* users confuse these terms with *grounding* and *grounded*. Although the terms are not the same, they are related concepts. See *NEC* Article 250 for more information.

Bonding Jumper, Main. The connection between the grounded circuit conductor and the equipment grounding conductor at the service.

The main bonding jumper (MBJ) is installed at the service equipment, and may be a wire, busbar, or screw; see 250.28 for more information about main bonding jumpers. In electrical installations covered by the *NEC*, the MBJ is the only connection permitted between the grounded conductor system (conductors with white or gray insulation, which sometimes are neutrals) and the grounding conductor system (equipment grounding conductors and grounding electrode conductors; these may be either bare or have a predominantly green insulation or covering).

Branch Circuit. The circuit conductors between the final overcurrent device protecting the circuit and the outlet(s).

Branch circuit does not include utilization equipment supplied by an outlet, devices such as receptacles and wall switches installed at outlets, or power supply cords with attachment plugs of utilization equipment plugged into receptacles.

Branch Circuit, Individual. A branch circuit that supplies only one utilization equipment.

A single receptacle installed on an individual branch circuit must have an ampere rating not less than that of the branch circuit [210.21(B)(1)], so a 20-ampere single receptacle cannot be installed on a 15-ampere individual branch circuit (**Figure 2.10**). A duplex receptacle is actually two receptacles; see the definition of receptacle. An older term for *individual branch circuit* is *dedicated circuit*; however, this term is not defined in the *Code*.

Branch Circuit, Multiwire. A branch circuit that consists of two or more ungrounded conductors that have a voltage between them, and a grounded conductor that has equal voltage between it and each ungrounded conductor of the circuit and that is connected to the neutral or grounded conductor of the system.

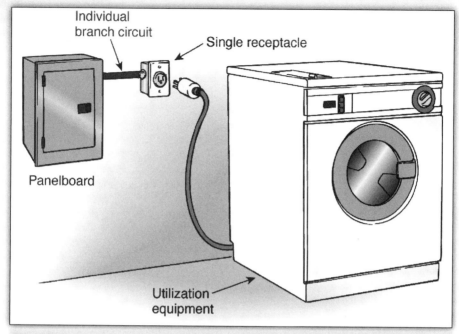

Figure 2.10 An individual branch circuit supplies only one piece of utilization equipment. Only a single receptacle is permitted on an individual branch circuit that supplies cord-and-plug–connected equipment. Reprinted with permission from 2005 NEC Handbook, copyright © 2005, National Fire Protection Association.

Multiwire branch circuits are frequently used to economize on homerun wiring by sharing a grounded (neutral) conductor among either two or three phase conductors. Multiwire branch circuits also are used for applications such as serving duplex receptacles on kitchen counters or workbenches, where the possibility of plugging in high-load appliances makes it desirable to wire alternate receptacles on different circuits. See 210.4, 240.20(B)(1), and 300.13(B) for additional information about multiwire branch circuits.

Building. A structure that stands alone or that is cut off from adjoining structure by fire walls with all openings therein protected by approved fire doors.

According to this definition, parts of a common structure separated by fire-rated partitions, such as warehouse spaces and townhouses, are considered to be separate buildings even if physically joined together. Classification of buildings and structures is not within the scope of the *NEC*, so see NFPA 5000, *Building Construction and Safety Code*.

Continuous Load. A load where the maximum current is expected to continue for 3 hours or more.

What constitutes a continuous load often is a matter for discussion with the AHJ; see 210.20(A) and 215.3. Automatically controlled lighting in a commercial structure that operates for 10 or 12 hours daily is a continuous load. Wall-switch controlled lighting in a dwelling unit may not be. The *Code* specifies which loads must be considered continuous in only a couple of places.

426.4 Continuous Load
Fixed outdoor electric deicing and snow-melting equipment shall be considered as a continuous load.

427.4 Continuous Load
Fixed electric heating equipment for pipelines and vessels shall be considered continuous load.

Demand Factor. The ratio of the maximum demand of a system, or part of a system, to the total connected load of a system or the part of the system under consideration.

Device. A unit of an electrical system that carries or controls energy as its principal function.

Devices include components such as switches, dimmers, receptacles, lampholders, circuit breakers, fuses, and terminal blocks that distribute control of electricity but do not consume it. Components that consume electricity, such as motors, heaters, and luminaires **(Figure 2.11)**, are considered utilization equipment (see the following definition). Wall switches, dimmers, and receptacles are commonly called *wiring devices;* however, this term isn't defined in the *Code.*

Disconnecting Means. A device, group of devices, or other means by which the conductors of a circuit can be disconnected from their source of supply.

Many different devices are permitted to serve as required disconnecting means for different types of equipment, including wall switches and attachment plugs under certain circumstances. Circuit breakers are considered disconnecting means. Fuses by themselves aren't considered disconnecting means; however, a fused switch or pullout block that simultaneously disconnects all ungrounded (phase) conductors of a circuit is considered a disconnecting means.

The common field terms *disconnect switch* and *safety switch* aren't used in the *NEC.*

Dwelling Unit. A single unit, providing complete and independent living facilities for one or more persons, including permanent provisions for living, sleeping, cooking, and sanitation.

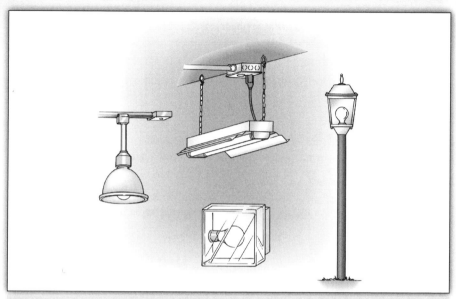

Figure 2.11 A luminaire is a complete lighting unit that includes the lamp, lens or diffuser, ballast, and other components.

An apartment-type hotel or motel room with its own kitchen area meets the definition of *dwelling unit*. Hotel and motel guest rooms without permanent cooking arrangements aren't considered dwelling units.

Enclosure. The case or housing of apparatus, or the fence or walls surrounding an installation to prevent personnel from accidentally contacting energized parts or to protect the equipment from physical damage.

Box-type enclosures are normally classified according to a NEMA/UL numbering system that rates them for protection against environmental conditions, as shown in *NEC* Table 110.20. Three of the most common NEMA/UL electrical enclosure types are:

1. 1: Indoor use, dry locations
2. 3R: Indoor or outdoor use, wet locations, "rainproof"
3. 4: Indoor or outdoor use, wet locations, "watertight"

Note

The *National Electrical Code* doesn't use the IP (ingress protection) enclosure classification system defined in International Electrotechnical Commission (IEC) standards.

Type 4 enclosures are normally used where equipment is exposed to hose-directed waters (for example, car washes, dairies, and industrial locations that are washed down).

Exposed (as applied to wiring methods). On or attached to the surface or behind panels designed to allow access.

Exposed wiring methods include those located behind access doors or above a dropped ceiling with lift-out panels.

Feeder. All circuit conductors between the service equipment, the source of a separately derived system, or other power and the final branch-circuit overcurrent device.

Grounding conductors are not intended to carry current under normal circumstances. They are bare or have predominantly green insulation, and are covered by Article 250.

Ground-Fault Circuit Interrupter. A device intended for the protection of personnel that functions to de-energize a circuit or portion thereof within an established period of time when a current to ground exceeds the values established for a Class A device.

This term refers to a device designed to protect people against electrocution. A GFCI measures the current in both conductors of a two-wire circuit and automatically shuts off power when it detects a current imbalance of 4–6 milliamperes (mA) between two circuit conductors. All GFCIs manufactured today are "Class A" devices; an older type of GFCI known as Class B is no longer manufactured.

Identified (as applied to equipment). Recognizable as suitable for the specific purpose, function, use, environment, application, and so forth, where described in a particular *Code* requirement.

This is not the same as listing, although frequently it is related. Examples of "identified" are a luminaire marked as being suitable for wet locations, and an enclosure marked for use in Class II, Division 2 hazardous (classified) locations.

In Sight From (Within Sight From, Within Sight). Where this *Code* specifies that one equipment shall be "in sight from," "within sight from," or "within sight," and so forth, of another equipment, the specified equipment is to be visible and not more than 15 m (50 ft) distant from the other.

This definition typically applies to utilization equipment and disconnecting means. Under many circumstances, a disconnect is required to be within sight of and not more than 15 m (50 ft) from its associated equipment. The intent is to make sure that a person maintaining or repairing that utilization equipment can verify that the disconnecting means is off and the equipment is deenergized.

Isolated (as applied to location). Not readily accessible to persons unless special means for access are used.

High-bay luminaires (lighting fixtures) that can be relamped or repaired only by using scaffolding or power lifts are considered *isolated*.

Labeled. Equipment or materials to which has been attached a label, symbol, or other identifying mark of an organization that is acceptable to the authority having jurisdiction and concerned with product evaluation, that maintains periodic inspection of production of labeled equipment or materials, and by whose labeling the manufacturer indicates compliance with appropriate standards or performance in a specified manner.

This is not the same as *listing*, although frequently it is related. Most listed equipment also is labeled to identify the listing organization, and in some cases the permitted conditions of use. In the case of small products such as conduit fittings and wire connectors, the labeling may appear on the smallest unit of packaging and/or in the manufacturer's instructions. Also see *identified*.

Listed. Equipment, materials, or services included in a list published by an organization that is acceptable to the authority having jurisdiction and concerned with evaluation of products or services, that maintains periodic inspection of production of listed equipment or materials or periodic evaluation of services, and whose listing states that the equipment, material, or services either meets appropriate designated standards or has been tested and found suitable for a specified purpose.

This definition summarizes the way that product listing agencies work. Product listing provides the basis for AHJ approval of products; however, it is important to understand that listing agencies themselves don't approve products. Many, but not all, electrical products covered by the *NEC* are listed; products not usually listed include motors, lamps, some types of cables, and custom one-of-a kind or imported apparatus.

Premises Wiring (System). Interior and exterior wiring, including power, lighting, control, and signal circuit wiring together with all their associated hardware, fittings, and wiring devices, both permanently and temporarily installed. This includes (a) wiring from the service point or power source to the outlets or (b) wiring from and including the power source to the outlets where there is no service point.

Such wiring does not include wiring internal to appliances, luminaires, motors, controllers, motor control centers, and similar equipment.

This general term covers all power, communications, and control wiring of a building or similar structure (including fiber optic systems), from the service point or other source to the outlets.

Qualified Person. One who has skills and knowledge related to the construction and operation of the electrical equipment and installations and has received safety training to recognize and avoid the hazards involved.

In general, NFPA 70E and the *National Electrical Code* define what tasks must be performed and what precautions must be taken to create a safe installation, but does not specify who shall perform them. Licensed electricians frequently perform construction and maintenance work covered by *Code* requirements, but not all jurisdictions have electrician licensing. Engineers, technicians, and specialized installers perform some types of work described in the *NEC*, and many jurisdictions permit property owners to perform electrical work on their own property. For these reasons, Article 100 includes a definition of *qualified person*, and this term is used many places throughout the *Code* to indicate competency.

 Separately Derived System. A premises wiring system whose power is derived with a means of sealing or locking so that live parts cannot be made accessible without opening the enclosure. The equipment may or may not be operable without opening the enclosure.

Typically, a *separately derived system* originates at a generator or transformer, but it also can be supplied from sources such as solar photovoltaic systems or fuel cells. Most emergency and optional standby power systems are also *separately derived systems*.

 Service. The conductors and equipment for delivering electric energy from the serving utility to the wiring system of the premises served.

Conductors and equipment that subfeed energy from another building or structure, or supply it from a separately derived system, are not considered services even if they are the only source of electrical energy for a building or structure.

    Conductors Associated with Services. The *National Electrical Code* uses a number of confusingly similar terms to describe conductors associated with services. Defining them by the order in which they physically occur is clearer than by the alphabetical arrangement used in Article 100 (Figures NECH 100.13 and 100.14).

    Service Drop (or Service Lateral) Conductors. These conductors are owned by the utility and connect the utility's network to the service point. *Service drop* conductors run overhead. *Service lateral* conductors run underground. Both are outside the scope of the *NEC*.

> **Note**
>
> Where service equipment is located outside the building walls, there may be no service-entrance conductors or they may be entirely outside the building.

    Service Conductors. The conductors from the service point to the service disconnecting means. They are within the scope of the *NEC*.

    Service-Entrance Conductors (Overhead System or Underground System). The service conductors between the terminals of the service equipment and the point of connection to the service drop (for an overhead system) or the service lateral (for an underground system).

Service Point. The point of connection between the facilities of the serving utility and the premises wiring.

This definition is a crucial *National Electrical Code* concept. Conductors and equipment on the customer side (load side, downstream) of the service point are covered by *NEC* rules. Conductors and equipment on the utility side (line side, supply side, upstream) of the service point aren't covered by the *NEC*. Typically, these are constructed according to the *National Electrical Safety Code*, or the serving utility's own rules and practices (**Figures 2.12** and **2.13**).

However, as previously discussed in the commentary on 90.2 Scope, normal lighting and power installations in buildings, yards, and rights-of-way owned and used by utilities are covered by the *NEC*. Roadway, area, and parking lot lighting installations operated by electric utilities for municipalities or private customers also are covered by the *NEC*. The only portions of utility-owned systems and equipment that are outside the scope of the *Code* are those portions directly related to the core activity of the utility: distributing electrical energy or communications services.

Special Permission. The written consent of the authority having jurisdiction.

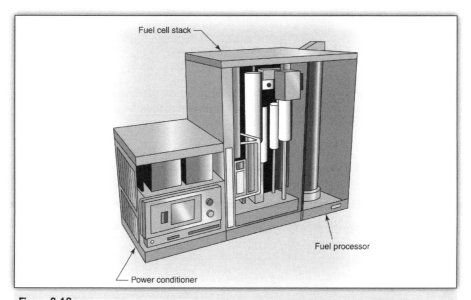

Figure 2.12 A residential fuel cell assembly is one type of separately derived system. Reprinted with permission from 2005 NEC Handbook, copyright © 2005, National Fire Protection Association.

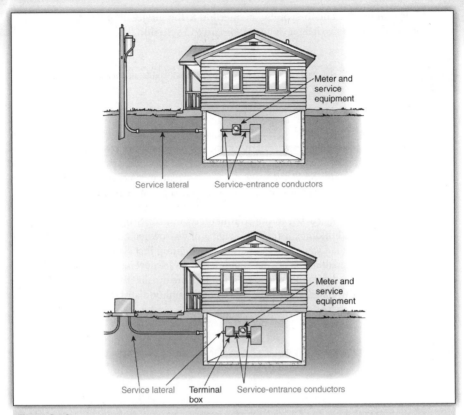

Figure 2.13 Underground systems are supplied by service laterals from pole- or pad-mounted transformers.

This term is used in 90.4, which allows the AHJ to waive particular requirements or to approve alternate construction methods. Special permission must be given in writing. Nowadays, this is normally interpreted to include e-mail.

Utilization Equipment. Equipment that utilizes electric energy for electronic, electromechanical, chemical, heating, lighting, or similar purposes (**Figure 2.14**).

Compare this definition of utilization equipment to the definition of *device*.

Voltage, Nominal. A nominal value assigned to a circuit or system for the purpose of conveniently designating its voltage class (e.g., 120/240 volts, 480Y/277 volts, 600 volts). The actual voltage at which a circuit operates can vary from the nominal within a range that permits satisfactory operation of equipment.

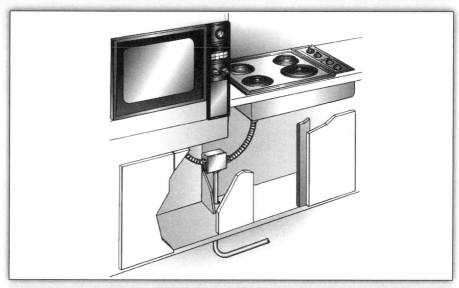

Figure 2.14 Cooking appliances are one familiar type of utilization equipment. Reprinted with permission from 2005 NEC Handbook, copyright © 2005, National Fire Protection Association.

The *NEC* uses multiples of 120 volts to describe system and circuit voltages [210.6, 310.15(B)(6)] and multiples of 125 volts to describe equipment voltage ratings [210.8(A)]. System and circuit voltages follow ANSI C84.1, *Electric Power Systems and Equipment— Voltage Ratings (60 Hz)*, and equipment voltage ratings are based on UL product listing standards.

Most installations covered by the *NEC* operate at 600 volts and less, and most electrical construction products are rated and/or listed for use at a maximum of 600 volts. The reason for this is that Canada uses 347/600 volts, 3-phase, 4-wire systems. Rating products at a maximum of 600 volts means that many of the same electrical construction products can be used throughout the North American market.

Voltage to Ground. For grounded circuits, the voltage between the given conductor and that point or conductor of the circuit that is grounded; for ungrounded circuits, the greatest voltage between the given conductor and any other conductor of the circuit.

Terms Not Defined in Article 100

Article 100 contains definitions of general terms used throughout the *Code*. Other, more specialized terms used in a single article are defined in those same articles, normally in the second numbered section. In addition, there are terms not formally defined in the *NEC* that sometimes confuse users. A noninclusive list of useful *Code* terms not defined in Article 100 follows.

Arc-fault circuit interrupter (AFCI). An arc-fault circuit interrupter is a device intended to provide protection from the effects of arc faults by recognizing characteristics unique to arcing and by functioning to deenergize the circuit when an arc fault is detected.

This definition appears in 210.12. AFCIs protect against low-energy arcing faults that may not be of sufficient magnitude to activate a branch-circuit overcurrent protection device, but could ignite flammable materials such as paper and fabric.

Cable. A cable is a factory assembly of two or more conductors having an overall covering (**Figure 2.15**).

Although this term is widely used throughout the *NEC*, the definition appears in 800.2.

Cable sheath. A cable sheath is a covering over the conductor assembly that may include one or more metallic members, strength members, or jackets. Although this term is widely used throughout the *NEC*, this definition appears in 800.2.

Supplemental grounding electrode. A metal underground water pipe used as a grounding electrode must be supplemented by an additional electrode [250.53(D)(2)]. This mandatory additional electrode is called a *supplemental grounding electrode.*

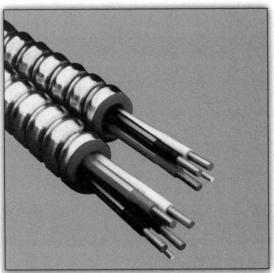

Figure 2.15 Type AC cable includes insulated conductors and a bare aluminum grounding strip.

Tap conductor. A conductor, other than a service conductor, that has overcurrent protection ahead of its point of supply that exceeds the value permitted for similar conductors by 240.4 [240.2].

Voltage ratings, straight and slash. NEC 240.85 gives 240V or 480V as examples of straight voltage ratings, and 120/240V or 480Y/277V as examples of slash voltage ratings.

Wire. A factory assembly of one or more insulated conductors without an overall covering. Although this term is widely used throughout the *NEC*, this definition appears in 800.2.

Units of Measurement

The *National Electrical Code* is dual dimensioned, using both the modernized metric system known as the International System of Units (SI) and traditional English or inch-pound units. The SI units appear first, followed by inch-pound units in parentheses [90.9]. Compliance with the numbers shown in either the SI system or the inch-pound system constitutes compliance with the *Code* [90.9(D)].

Rationale

The *NEC* is dual dimensioned to comply with the *NFPA Style Manual* and to improve its acceptability as an international standard. However, at this time, the North American construction market uses inch-pound units exclusively.

How the *NEC* Is Related to International Standards

NFPA fire protection standards are widely used around the globe, and many have been adopted by the International Organization for Standardization (ISO). However, the *National Electrical Code* isn't closely related to international standards for installing electrical products and systems, primarily because North American electrical systems operate at different voltages and frequencies from those in many other parts of the world.

International Electrotechnical Commission (IEC) Standards

Although the *NEC* is a true international standard, translated into other languages and used in other countries, it bears little resemblance to IEC 60364-1, *Electrical Installations of Buildings*. The *Code* contains many coordinated requirements in a single volume. By contrast, IEC 60364 is a collection of standards that are revised at different times; some of the standards in the series have very long revision cycles.

IEC performance standards. In general, IEC standards are performance-based documents with flexible requirements that allow many details to be determined by the design engineer's judgment. The closest parallel in the *Code* is probably 310.15(C), which permits conductor ampacities to be calculated under engineering supervision, using a formula commonly known as the Neher-McGrath Method.

U.S. prescriptive standards. By contrast, the *NEC* is a more prescriptive document that specifies more details of electrical installations, such as equipment ratings, conductor ampacities, raceway sizes, working space about electrical equipment, clearances from live parts, and installation techniques. The same is true of other U.S. codes and standards intended for regulatory adoption and use. Being prescriptive, and thus more uniform, makes them more suitable for regulatory enforcement because all users (owners, design engineers, installers, and inspectors) operate from the same set of shared assumptions and expectations.

Equivalent levels of safety. Thus, 90.1(D) should not be read as meaning that the NFPA 70 and IEC 60364-1, Section 131 have equivalent technical requirements. Rather, it means that electrical systems designed in accordance with both documents adhere to the same general principles and offer equivalent levels of safety [90.1(D) FPN].

Canadian Electrical Code

In the same way, designing and installing electrical systems to the Canadian Electrical Code (CEC) produces a level of safety similar to what is achieved following the *NEC*. Many electrical products are cross-listed for use in both countries, which permits them to be sold throughout the North American market.

However, the two nations' regulatory electrical codes are not compatible with each other. They differ in significant ways, including terminology, voltage ranges, cable and conductor types, wiring device configurations, and reliance on a wholly different set of product safety standards. Canadian electrical products are listed to standards of CSA International (formerly the Canadian Standards Association), while U.S electrical products are listed to the standards shown in *NEC* Annex A (primarily ANSI/UL standards).

Conclusion

This unit discusses general concepts that engineers must understand to use the *National Electrical Code* as a tool in the electrical design process, including the importance of Article 100 definitions. Much of it deals with how NFPA 70 is adopted and enforced by state and local governments as their official wiring rules, how product listing works in conjunction with the *NEC* wiring rules, and an explanation of why the UL *Guide Information for Electrical Equipment (White Book)* is an important reference for electrical engineers and designers.

UNIT 3

Conductors, Circuits, and Wiring Methods

Introduction

The first two units of this book covered *National Electrical Code* structure, organization, and language, along with general concepts that influence how engineers and other designers use the *Code* as part of their design process. This is the first of four units that explain how actual *NEC* rules affect the way in which engineers make design decisions and choose among competing methods and technologies.

The purpose of NFPA 70 is the "practical safeguarding of persons and property from hazards rising from the use of electricity" [90.1]. One of its primary concerns is preventing conductors from overheating; 310.10 states: "No conductor shall be used in such a manner that its operating temperature exceeds that designated for the type of insulated conductor involved."

Safety rules governing circuit types, conductors, and wiring methods are located throughout the *Code*, but concentrated in Chapters 2 and 3. This unit covers the following:

- Conductor Materials and Sizes
- Circuit Types
- Conductor Identification
- Equipment Terminals
- Conductor Ampacities
- Overcurrent Protection
- Load Calculations
- Harmonics Caused by Nonlinear Loads
- Voltage Drop
- Wiring Methods
- Organization of *NEC* Chapter 3
- Cable and Raceway Articles
- Determining Raceway Sizes

Conductor Materials and Sizes

Conductor Materials

All conductors in the *NEC*, whether wires or busbars, are assumed to be copper unless specifically identified as another material [110.5]. Aluminum and copper-clad aluminum conductors are widely used for general applications within the scope of the *Code*; other materials are occasionally used in specialized applications [Table 310.1, 810.11]. Aluminum and copper-clad aluminum have identical ampacities and other characteristics.

Conductor Sizes

Units. Conductor sizes are expressed as American Wire Gauge (AWG) in sizes 18 AWG through 4/0 AWG. (Older terminology to be avoided is *No. 18* and *#18*, *No. 4/0* and *#4/0*.) Larger conductors are designated in thousand circular mils for sizes 250 kcmil through 2000 kcmil (avoid the older usage *MCM*).

SI units. Although the *Code* is generally dual dimensioned, conductor sizes aren't expressed in SI units because conductors in metric sizes aren't readily available in North America [90.9(C)(3)]. The reference tables in Chapter 9 include metric dimensions for conductor and raceway and conductor sizes, but *NEC* Chapters 1–8 don't refer to conductor sizes using metric units.

Solid and stranded conductors. Conductors 8 AWG and larger must be stranded when installed in raceways [310.3]. However, many engineers prefer to specify stranded 10 AWG conductors as well, because they are easier to pull.

Circuit Types

Conductors for permanent installation under *National Electrical Code* rules generally fall into one of four types of circuits (**Figure 3.1**). Definitions for some of these terms were included in the list at the end of Unit 2; they are repeated here for convenience:

- Service conductors
- Feeders
- Branch circuits
- Tap conductors

 Service conductors. The conductors from the service point to the service disconnecting means.

Service conductors are within the scope of Article 230.

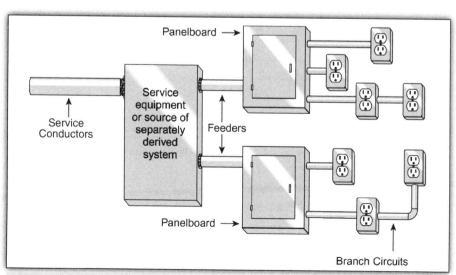

Figure 3.1 Service, feeder, and branch-circuit conductors. Reprinted with permission from 2005 NEC Handbook, copyright © 2005, National Fire Protection Association.

Feeder. All circuit conductors between the service equipment, the source of a separately derived system, or other power and the final branch-circuit overcurrent device.

Rules for feeders appear in Article 215.

 Branch Circuit. The circuit conductors between the final overcurrent device protecting the circuit and the outlet(s).

Rules for branch circuits appear in Article 210. There are a number of different kinds of branch circuits, as defined in Article 100; and different *Code* rules apply to them. Engineers should pay particular attention to multiwire branch circuits and individual branch circuits.

 Tap Conductor. A conductor, other than a service conductor, that has overcurrent protection ahead of its point of supply that exceeds the value permitted for similar conductors by 240.4.

Taps are not actually circuits in the same sense as the other three. Instead, they are a way of connecting smaller conductors directly to larger ones, and are used in various applications. For example, a common design for services of strip shopping centers, two-family dwellings, and other types of buildings is to run the service-entrance conductors from the utility into a meter enclosure or wireway (often called *wiring trough* in the field), feeding multiple panelboards or disconnect switches (**Figure 3.2**). The conductors from the main service conductors to each individual piece of service equipment are tap conductors. Because they are used in many different ways, rules for tap conductors appear throughout the *NEC*:

- Branch circuits—210.19(A), Ex. No. 1
- Busways—368.17(B) and (C), Ex. 1
- Feeders—240.21(B), 430.28
- Fixture wire—240.5(B)(2)
- Grounding electrode conductors—250.64(D)
- Household ranges and cooking equipment—210.19(A)(3) and (4)
- Motor taps—430.53(D)
- Overcurrent protection—240.21
- Remote-control circuits—725.24(B) and (C)
- Service-entrance conductors—230.46

Other Types of Wiring of Equipment

These four types of wiring are covered by the rules of *NEC* Chapters 2 and 3. Other types of wiring covered by different *Code* chapters are discussed further in the section on wiring methods, later in this unit.

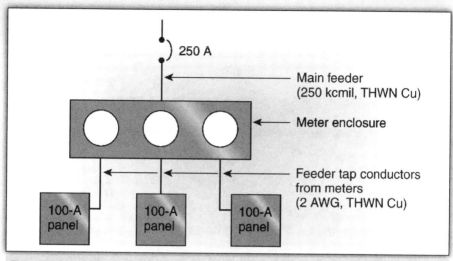

Figure 3.2 Three sets of tap conductors are tapped from the incoming service-lateral conductors to supply three panelboards. As shown, these tap conductors are also service-entrance conductors and feeder conductors, depending upon their location in the premises wiring system. Reprinted with permission from 2005 NEC Handbook, copyright © 2005, National Fire Protection Association.

Conductor Identification

No Color Code for Phase Conductors

The *National Electrical Code* doesn't specify insulation colors for ungrounded (phase) conductors. As a safety code, rather than a design specification, it specifies insulation colors only where identifying a particular conductor is critical for safety. The following ungrounded (phase) conductor color codes (**Table 3.1**) are commonly included in project specifications, but they don't appear anywhere in the *NEC*.

However, the *Code* does have rules for identifying several classes of conductors where this identification is particularly important for safety. These classes are:

- Grounded (Neutral) Conductors
- Grounding Conductors
- High-Leg Conductors

Table 3.1 Ungrounded Conductor Color Codes

Phase	208Y/120 Volts 3-phase, 4-wire	480Y/277 Volts 3-phase, 4-wire	120/240 Volts single-phase, 3-wire
A	Black	Yellow	Black
B	Red	Orange	Red
C	Blue	Brown	—
Grounded (neutral)	White	Gray	White

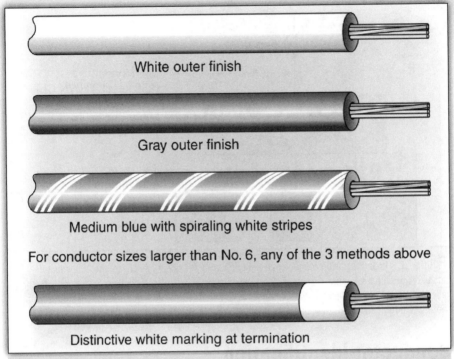

White outer finish

Gray outer finish

Medium blue with spiraling white stripes

For conductor sizes larger than No. 6, any of the 3 methods above

Distinctive white marking at termination

Figure 3.3 Most conductors 6 AWG and larger have black insulation. White tape or paint is typically applied near conductor terminations in the field to identify grounded (neutral) conductors.

Grounded (Neutral) Conductors

Article 200 specifies means to identify these conductors and the electrical equipment terminals to which they are connected. Grounded (neutral) conductors must have insulation that is predominantly white or gray, or have white or gray marking at terminals [200.6, 200.7]. White tape or paint is typically applied in the field to conductors 6 AWG and larger (**Figure 3.3**).

NEC 400.22 and 402.8 have similar identification requirements for grounded (neutral) conductors used in flexible cords, cables, and fixture wires.

NEC 200.9 requires that equipment terminals for grounded (neutral) conductors be substantially white in color, and that other terminals be "of a readily distinguishable different color." Electrical equipment terminals are normally silver colored for grounded (neutral) conductors, and brass or copper colored for ungrounded (phase) conductors.

Grounding Conductors

What are commonly called *grounding conductors* actually include two different types of conductors:

- *Equipment grounding conductors*, which are run with circuit conductors
- *Grounding electrode conductors*, which are connected to grounding electrodes

Equipment grounding conductors. Equipment grounding conductors can be bare, covered, or insulated. Individually covered or insulated equipment grounding conductors must have a continuous outer finish that is substantially green in color [250.119].

Conductors larger than 6 AWG aren't generally available in colors other than black. Insulated or covered conductors larger than 6 AWG are permitted to be identified at each end and every point where the equipment grounding conductor is accessible by stripping off the insulation or covering, coloring the exposed insulation green, or using green tape or adhesive labels [250.119(A)(2)].

Wiring device terminals for equipment grounding conductors are green and hexagonal, or marked as specified in 250.126.

Grounding electrode conductors. The *Code* doesn't specify an insulation color for grounding electrode conductors; they are typically bare.

High-Leg Conductors

On a 3-phase, 4-wire delta-connected system where the midpoint of one phase winding is grounded, the phase conductor with the higher voltage-to-ground must be durably and permanently marked by an orange outer finish or other effective means (such as tagging). This identification must appear at each point where the connection is made, if the grounded (neutral) conductor is also present.

Note

High-leg conductors are also known as "wild-leg," "stinger-leg," or "bastard-leg."

Insulation Markings

Different conductor insulation types and markings are required for dry, damp, and wet locations; where exposed to direct sunlight; under corrosive conditions; in different operating temperature ranges; and for specific applications, such as wiring inside switchboards. Table 310.13 lists conductor insulation trade names, characteristics, and type designations (RHW-2, THHN, TW, USE, etc.). Letters are used to designate qualities of conductor/cable insulation (**Table 3.2**).

Equipment Terminals

Temperature Ratings

Equipment terminals rated 100 amperes or less are intended for use with conductors rated 60°C (140°F), unless otherwise listed and identified. Equipment terminals rated over 100 amperes are intended for use with conductors rated 75°C (167°F), unless otherwise listed and identified [110.14(C)(1)].

When conductors with higher insulation temperature ratings are used, their ampacities are limited to the figure that matches the terminal temperature rating. If a 90°C (194°F) conductor insulation such as XHHW is specified, the 60°C (140°F) or 75°C (167° F) ampacity columns in Tables 310.16–301.18 must still be used.

Table 3.2 Most Common Letters Designating Conductor/Cable Insulation

R	Rubber
T	Thermoplastic
M	Metal, mineral
N	Nylon
W	Suitable for wet locations
H	High temperature (usually 75°C)
HH	Higher temperature (usually 90°C)
X	Cross-linked polyethylene
O	Oil-resistant
S	Hard service
J	Junior (hard service)
U	Underground

Figure 3.4 Most electrical equipment terminals (lugs/wire connectors) are rated for 75°C or 60/75°C, Al/Cu or Cu-only conductors.

Terminal Marking

Equipment terminals. Electrical equipment terminals without special markings are for use with copper conductors only. Terminals marked AL/CU can be used with copper, aluminum, or copper-clad aluminum conductors. In practice, most electrical equipment terminals are rated 75°C and marked AL/CU (Figure 3.4).

Wiring device terminals. Wiring device terminals without special markings are for use with copper conductors only. Receptacles and switches (including dimmers) rated 15 and 20 amperes that are marked CO/ALR can be used with copper, aluminum, or copper-clad aluminum conductors. In practice, all modern wiring devices are marked CO/ALR; older ones may not be.

Conductor Ampacities

Table 310.16 is the most commonly used table in the *Code* for determining conductor ampacities. It provides ampacities for both single-phase and 3-phase circuits enclosed in raceways, cables, or buried together in the earth, operating at up to 2,000 volts, based on conductor size and temperature rating (**Table 3.3**).

Tables 310.17 through 310.86 specify conductor ampacities under other, more specialized conductor installation situations, such as single conductors run in free air.

Adjustment Factors

Section 310.15(B)(2) and Table 310.15(B)(2) provide a number of adjustment factors that modify the ampacities given in the tables.

Residential Services and Feeders

Table 310.15(B)(6) lists conductor types and sizes specifically for 120/240-volt, 3-wire, single-phase dwelling services and feeders (**Table 3.4**).

Determining Conductor Size

There are six steps in the process of selecting a circuit conductor size:

1. Load—Calculate the load in amperes to be served by a branch-circuit, service, or feeder [Article 220]. Load calculations are discussed further later in this unit.
2. Terminals—Determine temperature rating [110.14(C)(1)].
3. Ampacity—Select from Article 310 tables, using the column that matches the terminal temperature rating. Table 310.16 is the most commonly used ampacity table in the *Code.*
4. Ambient temperature rating—Apply correction factor where ambient differs from 30°C (86°F) [Bottom of Tables 310.16, 310.17, 310.18, 310.19, and 310.20].
5. Number of conductors—Where more than three current-carrying conductors are run together in a raceway or cable, apply the adjustment factor in 310.15(B)(2)(a).
6. Conductor size—Select a conductor size that satisfies the adjusted ampacity.

Overcurrent Protection

Conductor Ampacity

Section 240.4 requires that service, feeder, and branch-circuit conductors be protected against overcurrent at the ampacities specified in 310.15, unless these conductors fall into one of seven special classes. The sections listed here provide more information for each special class of conductors:

- (A) Power Loss Hazard
- (B) Devices Rated 800 Amperes or Less (next size up)
- (C) Devices Rated Over 800 Amperes (next size down)

Table 3.3 Allowable Ampacities of Insulated Conductors Rated 0 Through 2,000 Volts, 60°C Through 90°C (140°F Through 194°F), Not More than Three Current-Carrying Conductors in Raceway, Cable, or Earth (Directly Buried), Based on Ambient Temperature of 30°C (86°F)

	Temperature Rating of Conductor						
	60°C (140°F)	75°C (167°F)	90°C (194°F)	60°C (140°F)	75°C (167°F)	90°C (194°F)	
Size AWG or kcmil	Types TW, UF	Types RHW, THHW, THW, THWN, XHHW, USE, ZW	Types TBS, SA, SIS, FEP, FEPB, MI, RHH, RHW-2, THHN, THHW, THW-2, THWN-2, USE-2, XHH, XHHW, XHHW-2, ZW-2	Types TW, UF	Types RHW, THHW, THW, THWN, XHHW, USE	Types TBS, SA, SIS, THHN, THHW, THW-2, THWN-2, RHH, RHW-2, USE-2, XHH, XHHW, XHHW-2, ZW-2	Size AWG or kcmil
	COPPER			ALUMINUM OR COPPER-CLAD ALUMINUM			
18	—	—	14	—	—	—	—
16	—	—	18	—	—	—	—
14*	20	20	25	—	—	—	—
12*	25	25	30	20	20	25	12*
10*	30	35	40	25	30	35	10*
8	40	50	55	30	40	45	8
6	55	65	75	40	50	60	6
4	70	85	95	55	65	75	4
3	85	100	110	65	75	85	3
2	95	115	130	75	90	100	2
1	110	130	150	85	100	115	1
1/0	125	150	170	100	120	135	1/0
2/0	145	175	195	115	135	150	2/0
3/0	165	200	225	130	155	175	3/0
4/0	195	230	260	150	180	205	4/0
250	215	255	290	170	205	230	250
300	240	285	320	190	230	255	300
350	260	310	350	210	250	280	350
400	280	335	380	225	270	305	400
500	320	380	430	260	310	350	500
600	355	420	475	285	340	385	600
700	385	460	520	310	375	420	700
750	400	475	535	320	385	435	750
800	410	490	555	330	395	450	800
900	435	520	585	355	425	480	900
1000	455	545	615	375	445	500	1000
1250	495	590	665	405	485	545	1250
1500	520	625	705	435	520	585	1500
1750	545	650	735	455	545	615	1750
2000	560	665	750	470	560	630	2000

(continued)

Table 3.3 *Continued*

CORRECTION FACTORS							
Ambient Temp. (°C)	**For ambient temperatures other than 30°C (86°F), multiply the allowable ampacities shown above by the appropriate factor shown below.**						**Ambient Temp. (°F)**
21–25	1.08	1.05	1.04	1.08	1.05	1.04	70–77
26–30	1.00	1.00	1.00	1.00	1.00	1.00	78–86
31–35	0.91	0.94	0.96	0.91	0.94	0.96	87–95
36–40	0.82	0.88	0.91	0.82	0.88	0.91	96–104
41–45	0.71	0.82	0.87	0.71	0.82	0.87	105–113
46–50	0.58	0.75	0.82	0.58	0.75	0.82	114–122
51–55	0.41	0.67	0.76	0.41	0.67	0.76	123–131
56–60	—	0.58	0.71	—	0.58	0.71	132–140
61–70	—	0.33	0.58	—	0.33	0.58	141–158
71–80	—	—	0.41	—	—	0.41	159–176

* See 240.4(D). Reprinted with permission from 2005 NEC Handbook, copyright © 2005, National Fire Protection Association.

Table 3.4 Adjustment Factors for More than Three Current-Carrying Conductors in a Raceway or Cable

Number of Current-Carrying Conductors	Percent of Values in Tables 310.16 through 310.19 as Adjusted for Ambient Temperature if Necessary
4–6	80
7–9	70
10–20	50
21–30	45
31–40	40
41 and above	35

Reprinted with permission from 2005 NEC Handbook, copyright © 2005, National Fire Protection Association.

- (D) Small Conductors
- (E) Tap Conductors
- (F) Transformer Secondary Conductors
- (G) Specific Applications

Selecting Overcurrent Devices

Standard overcurrent device ratings are listed in 240.4(A) (**Table 3.5**). Section 240.4(B)(2), the "next size up" rule, governs selection of overcurrent devices for most general applications. It's

Table 3.5 Standard Ampere Ratings of Overcurrent Devices Specified by 240.4(A)

Circuit Breakers and Fuses			Additional Standard Fuse Ratings
15	110	700	1
20	125	800	3
25	150	1000	6
30	175	1200	10
35	200	1600	601
40	225	2000	
45	250	2500	
50	300	3000	
60	350	4000	
70	400	5000	
80	450	6000	
90	500		
100	600		

important to understand that the provisions of 240.4(B) don't modify or change allowable conductor ampacity; they only provide a safe and acceptable degree of flexibility where conductor ampacities don't perfectly match the standard overcurrent device ratings.

For example, a 500-kcmil THWN conductor has an allowable ampacity of 380 amperes from Table 310.16. Section 240.4(B) permits this conductor to be protected by a 400-ampere overcurrent device rated 10 percent higher.

Continuous and Noncontinuous Loads

Article 100 defines a continuous load as one "where the maximum current is expected to continue for 3 hours or more." Four *Code* sections require that conductors' overcurrent devices (circuit breakers or fuses) not be sized less than 100 percent of the noncontinuous load, plus 125 percent of the continuous load (**Figure 3.5**).

- 210.19(A)—Branch circuit conductors
- 210.20(A)—Branch circuit overcurrent protective devices
- 215.2(A)—Feeder conductors
- 215.3—Feeder overcurrent protective devices

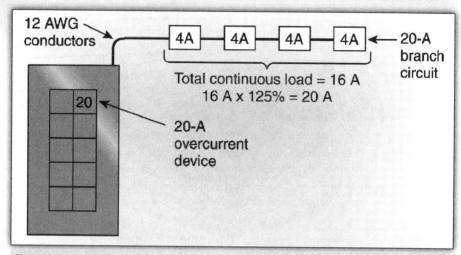

Figure 3.5 A continuous load (store lighting) is multiplied by 125 percent to determine the conductor ampacity and overcurrent-device rating. Reprinted with permission from 2005 NEC Handbook, copyright © 2005, National Fire Protection Association.

This means that for all practical purposes, overcurrent protective devices are sized at 125 percent of the loads they protect—or alternately, nearly all conductor ampacities are selected at 80 percent of the overcurrent device rating.

100 percent loading. Circuit breakers and fuses are generally listed to carry only 80 percent of their nameplate rating for 3 hours or more. Some overcurrent devices are listed to carry 100 percent of their nameplate rating on a continuous basis. However, these devices are generally available only in ratings of 400 amperes and higher.

Conductor and overcurrent device selection. Determine the minimum size THWN conductor, and minimum circuit breaker or fuse rating, to serve utilization equipment representing a 25-ampere continuous load.

1. 25 amperes × 1.25 = 31.25 amperes.
2. Next size up overcurrent device from 240.6 is 35 amperes.
3. Temperature rating isn't marked on equipment terminals, so assume 60°C per 110.14(C)(1).
4. Table 310.16 indicates ampacity of 35 amperes for 10 AWG THWN conductors.
5. However, the conductor temperature rating cannot exceed the terminal temperature rating. Using the 60°C column of Table 310-16, use 8 AWG conductors rated 40 amperes to connect to the OCPD-rated 35 amperes.
6. The 60°C column of Table 310.16 shows only Type TW and UF conductors. Other types with higher temperature ratings can be used for this application, but only at a 60°C allowable ampacity.

> **Note**
>
> For all practical purposes, most electrical equipment is designed to accept conductors sized to the 75°C column of Table 310.16.

Maximum branch-circuit continuous loads. The minimum circuit conductor size permitted by the *NEC* is 14 AWG, protected by a 15-ampere circuit breaker or fuse. The minimum branch circuit conductor size used in most nonresidential applications is 12 AWG, protected by a 20-ampere circuit breaker or fuse. Maximum loads for these common branch-circuit ratings are as follows:

- A 15-ampere branch circuit can support a maximum continuous load of 1440 volt-amperes at 120 volts. [15 amperes × 120 volts × 0.80 = 1440 volt-amperes]
- A 20-ampere branch circuit can support a maximum continuous load of 1920 volt-amperes at 120 volts. [20 amperes × 120 volts × 0.80 = 1920 volt-amperes]

> **Note**
>
> Strictly speaking, the term *ampacity* applies only to conductors. Equipment has ampere ratings or current ratings. However, in practice the term *ampacity* is frequently used to describe equipment's current characteristics.

Equipment Ratings

Section 240.3 requires equipment to be protected against overcurrent and lists the correct article for each type of equipment. This subject is covered in the next unit.

Load Calculations

Article 220 contains rules for calculating branch-circuit, feeder, and service loads; this is the first step in selecting conductor ampacities and overcurrent device ratings. Load calculation is surprisingly complex, and engineers should study the examples in Annex D of the *2005 National Electrical Code Handbook*.

Other Articles

Article 220 provides calculation methods that work satisfactorily for most common occupancies. Table 220.3 provides nearly 30 references to other *NEC* articles for more additional information about calculating loads for specialized occupancies and types of equipment.

Demand Factors

Article 100 defines these as: "The ratio of the maximum demand of a system, or part of a system, to the total connected load of a system or the part of the system under consideration." Demand factors are used in load calculations to account for the fact that not all utilization equipment installed in a building operates simultaneously [Tables 220.42, 220.44, 220.54, 220.55, 220.56, 220.84, 220.86, and Section 220.83].

Noncoincident Loads

Where it is unlikely that two or more loads will be used simultaneously, only the larger must be used when calculating the total load of a service or feeder [220.60]. This *Code* rule is typically applied when calculating loads in buildings that have both heating and air-conditioning systems.

Optional Calculations

Part IV of Article 220 describes optional feeder and service-load calculation methods for dwellings, schools, existing installations, and new restaurants. These optional calculations nearly always result in smaller feeder and service loads than those calculated using the "standard" methods in Parts I, II, and III of Article 220.

Neutral Loads

Grounded (neutral) conductors in two-wire feeders or services must be of the same size and ampacity as ungrounded (phase) conductors, to carry the return current. Under normal operating conditions, the *National Electrical Code* doesn't regard neutrals of 480Y/277- or 208Y/120-volt, 3-phase, 4-wire systems as current-carrying conductors, because the return currents on the different phases cancel each other out.

For this reason, 220.61 permits neutral conductors of feeders and services with three or more conductors to be reduced under various circumstances, often having to do with household cooking equipment. Examples D1(A), D1(B), D2(B), D4(A), and D5(A) in Annex D provide additional information. However, the growing phenomenon of nonlinear loads prevents reducing the size of neutral conductors in many commercial and industrial occupancies; this is explained next.

Harmonics Caused by Nonlinear Loads

Because they cause power-quality problems, nonlinear loads are very much in the (technical) news these days. A nonlinear load is one in which, due to the presence of harmonics, the waveform shape does not follow the waveform of the applied circuit voltage—typically a sine wave in AC power systems.

Nonlinear loads are a result of harmonic currents generated by electrical and electronic devices, including personal computers, laser printers, fluorescent and HID lighting ballasts, motor variable-speed drives, and uninterruptible power supplies (UPS). The rapid increase in the use of such equipment, particularly personal computers and electric discharge lighting in commercial and institutional buildings, has caused a great increase in harmonics due to nonlinear loads.

Harmonics caused by nonlinear loads affect both electrical power distribution systems and the utilization equipment supplied by them. One area of particular concern is the effect on grounded (neutral) conductors. Under normal operating conditions, the *Code* doesn't regard neutrals of 3-phase, 4-wire power distribution systems as current-carrying conductors. However, high levels of harmonics cause steady current on the neutral, in effect making it act more like a phase (ungrounded) conductor.

Code Provisions for Nonlinear Loads

Designing premises wiring systems to minimize unwanted harmonics is beyond the scope of this book. The next section summarizes *NEC* provisions dealing with this complex subject. For clarity, rules dealing with equipment and flexible cords and cables are also included in the following list:

Article 100 Nonlinear Load. A load where the wave shape of the steady state current does not follow the shape of the applied voltage.

FPN: Electronic equipment, electronic/electric-discharge lighting, adjustable speed-drive systems, and similar equipment may be nonlinear loads.

210.4 Multiwire Branch Circuits. The FPN to 210.4(A) cautions:

FPN: A 3-phase, 4-wire, wye-connected power system used to supply power to nonlinear loads may necessitate that the power system design allow for the possibility of high harmonic neutral currents.

220.61(C) Prohibited Reduction. Specifying grounded (neutral) conductor sizes smaller than the power (ungrounded) conductors is a common value-engineering technique in multiphase circuits with low neutral currents. However, the *NEC* prohibits this practice when nonlinear loads are involved.

310.4 Conductors in Parallel. Exception No. 4 to 310.4 can be used to alleviate overheating of neutral conductors in existing installations due to high harmonic currents. It permits installing an additional neutral run in parallel with the existing one.

310.15(B)(4)(c). This key *Code* provision defines the circumstances under which neutral conductors must be considered current-carrying conductors for purposes of raceway fill and derating. It reads:

On a 4-wire, 3-phase wye circuit where the major portion of the load consists of nonlinear loads, harmonic currents are present in the neutral conductor; the neutral shall therefore be considered to be a current-carrying conductor.

368.258 Neutral [of a busway]. Neutral bus, where required, shall be sized to carry all neutral load current, including harmonic currents, and shall have adequate momentary and short-circuit ratings consistent with system requirements.

400.5(B) Ultimate Insulation Temperature [of flexible cords and cables]. The fourth paragraph is very similar to 310.15(B)(4)(c):

On a 4-wire, 3-phase, wye circuit where the major portion of the load consists of nonlinear loads, there are harmonic currents present in the neutral conductor and the neutral shall be considered to be a current-carrying conductor.

450.3 [Transformer] Overcurrent Protection. FPN No. 2 warns that "Nonlinear loads can increase heat in a transformer without operating its overcurrent protective device."

450.9 [Transformer] Ventilation. FPN No. 2 contains the following warning:

FPN No. 2: Additional losses may occur in some transformers where nonsinusoidal currents are present, resulting in increased heat in the transformer above its rating. See ANSI/IEEE 110-1993, *Recommended Practice for Establishing Transformer Capability When Supplying Nonsinusoidal Loads,* where transformers are utilized with nonsinusoidal loads.

Annex D—Example D3(A) Industrial Feeders in a Common Raceway. In this feeder load calculation involving electric discharge lighting, the neutrals in two feeders are counted as current-carrying conductors.

Voltage Drop

The *NEC* doesn't have mandatory voltage drop rules for most general applications. However, for reasonable efficiency of operation, the FPNs to 210.19(A) and 215.2(A)(4) recommend that total voltage drop on both branch circuits and feeders be limited to 5 percent. These FPNs are informational and not enforceable. However, mandatory voltage drop requirements are provided in 647.4(D) for sensitive electronic equipment and 695.7 for fire pumps, where maintenance of nominal voltage is considered a safety, rather than a performance, requirement.

In addition, ASHRAE standard 90.1, *Energy Standard for Buildings Except Low-Rise Residential Buildings,* contains the following mandatory voltage drop requirements for all building power distribution systems:

- **8.2.1 Voltage Drop**
  - **8.2.1.1 Feeders.** Feeder conductors shall be sized for a maximum voltage drop of 2 percent at design load.
  - **8.2.1.2 Branch Circuits.** Branch circuit conductors shall be sized for a maximum voltage drop of 3 percent at design load.

Other regulatory standards. ASHRAE 90.1 is referenced in building codes such as NFPA 5000, *Building Construction and Safety Code,* and the *International Building Code.* Where these standards are adopted for regulatory use, the ASHRAE voltage drop requirements become mandatory. This subject is discussed further in Unit 7.

Wiring Methods

The general term *wiring method* covers several different techniques for installing conductors under the rules of *NEC* Chapters 1–4:

- Individual conductors installed in raceways
- Individual conductors installed within enclosures such as surface-mounted raceways, wireways, and multioutlet assemblies
- Cables containing two or more conductors
- Busways and cablebus
- Messenger-supported wiring and open wiring on insulators

All of these different techniques for installing conductors are considered Chapter 3 wiring methods. *NEC* Chapter 3 contains the primary rules for wiring methods, along with rules for installing enclosures, boxes, and cabinets used to produce a complete *premises wiring system,* as defined in Article 100.

Other Types of Wiring and Related Items

However, the following types of wiring and related items aren't considered Chapter 3 wiring methods. They are covered under different chapters of the *National Electrical Code.*

Flexible cords, cables, and fixture wires. Flexible cords and cables [Article 400] and fixture wires [Article 402] aren't considered wiring methods, and for this reason aren't included in Chapter 3. None of these cords, cables, and wires can be used as a substitute for fixed premises wiring or where concealed behind walls, floors, or ceilings [400.7, 402.11].

Internal wiring of equipment. In general, the *NEC* doesn't concern itself with internal wiring of equipment. Instead, detailed requirements appear in product safety standards (e.g., UL listing standards included in the Appendix to this book). The major exceptions are internal wiring of luminaires covered in Article 402, and internal wiring of motor controllers covered in Article 430.

Low-voltage and limited energy systems. Control, communications, and signaling conductors within the scope of *NEC* Chapters 7 and 8 aren't considered Chapter 3 wiring methods. *Code* rules for these low-voltage and limited-energy systems are described in Unit 5 of this book.

Cable trays. Although these are part of Chapter 3, cable trays aren't considered a wiring method. Instead, cables trays are a support system for cables, raceways, and enclosures.

Organization of *NEC* Chapter 3

Chapter 3 is divided into 44 articles covering major subjects. They can be considered in five broad categories: General, Cables, Raceways of Circular Cross-Section, Noncircular Raceways, and Other Wiring Methods and Support Systems (**Table 3.6**).

General Requirements for Wiring Methods

The first four Category 1 articles in *NEC* Chapter 3 can be considered as broad "horizontal" articles that apply across the other 40 narrow "vertical" articles, each of which is devoted to a specific type of cable, raceway, or other system. Articles 300, 310, 312, and 314 contain general rules that must be followed when using any wiring method, except when specifically modified in another article.

The general rules of Chapter 1 also apply to Chapter 3 wiring methods. Key requirement rules that engineers must be aware of when dealing with Chapter 3 wiring methods include the following:

Accumulated wires and cables [300.23]. Cables and raceways installed behind panels designed to allow access (such as suspended ceiling panels) must be arranged to allow the removal of panels and access to equipment. This rule is intended to prevent the accumulation of loose wires and cables above suspended ceilings. This subject also is dealt with in Chapters 7 and 8.

Bonding and grounding. Metal cable armor, metal raceways, and metal boxes must be joined together so they are electrically continuous [300.10]. Metal raceway systems must be grounded.

Boxes, conduit bodies, and fittings. A box, conduit body, or fitting must be installed at each outlet, switch location, or conductor splice point [300.15]. These boxes, enclosures, and fittings must have covers [300.31] and be accessible [314.29]. Splices are prohibited within raceways [300.13] (**Figure 3.6**).

Box sizing. Section 314.28 provides rules for calculating required sizes of pull boxes, junction boxes, and conduit boxes, based on the number of conductors enclosed.

Combustible materials [314.20]. When recessed into wood and other combustible wall constructions, cabinets, cutout boxes, and meter enclosures must be mounted so that the front edge is flush with or projects from the surface to prevent sparks from arcing or malfunctioning devices from igniting combustible materials.

Complete raceway and cable installations. Raceway and cable armor or sheaths must be continuous between boxes, cabinets, or other enclosures [300.12]. Raceway systems must be completely installed before conductors are pulled into them [300.18].

Table 3.6 Chapter 3 Categories of Articles

Category 1 Articles—General	Article 300—Wiring Methods Article 310—Conductors for General Wiring Article 312—Cabinets, Cutout Boxes, and Meter Socket Enclosures Article 314—Outlet, Device, Pull, and Junction Boxes; Conduit Bodies; Fittings; and Manholes
Category 2 Articles—Cables	Article 320—Armored Cable: Type AC Article 322—Flat Cable Assemblies: Type FC Article 324—Flat Conductor Cable: Type FCC Article 326—Integrated Gas Spacer Cable: Type IGS Article 328—Medium Voltage Cable: Type MV Article 330—Metal-Clad Cable: Type MC Article 332—Mineral-Insulated, Metal-Sheathed Cable: Type MI Article 334—Nonmetallic-Sheathed Cable: Types NM, NMC, and NMS Article 336—Power and Control Tray Cable: Type TC Article 338—Service-Entrance Cable: Types SE and USE Article 340—Underground Feeder and Branch-Circuit Cable: Type UF
Category 3 Articles—Raceways of Circular Cross-Section	Article 342—Intermediate Metal Conduit: Type IMC Article 344—Rigid Metal Conduit: Type RMC Article 348—Flexible Metal Conduit: Type FMC Article 350—Liquidtight Flexible Metal Conduit: Type LFMC Article 352—Rigid Polyvinyl Chloride Conduit: Type PVC Article 353—High-Density Polyethylene Conduit: Type HDPE Article 354—Nonmetallic Underground Conduit with Conductors: Type NUCC Article 355—Reinforced Thermosetting Resin Conduit: Type RTRC Article 356—Liquidtight Flexible Nonmetallic Conduit: Type LFNC Article 358—Electrical Metallic Tubing: Type EMT Article 360—Flexible Metallic Tubing: Type FMT Article 362—Electrical Nonmetallic Tubing: Type ENT
Category 4 Articles—Noncircular Raceways	Article 366—Auxiliary Gutters Article 368—Busways Article 370—Cablebus Article 372—Cellular Concrete Floor Raceways Article 374—Cellular Metal Floor Raceways Article 376—Metal Wireways Article 378—Nonmetallic Wireways Article 380—Multioutlet Assembly Article 382—Nonmetallic Extensions Article 384—Strut-Type Channel Raceway Article 386—Surface Metal Raceways Article 388—Surface Nonmetallic Raceways Article 390—Underfloor Raceways
Category 5 Articles—Other Wiring Methods and Support Systems	Article 392—Cable Trays Article 394—Concealed Knob-and-Tube Wiring Article 396—Messenger Supported Wiring Article 398—Open Wiring on Insulators

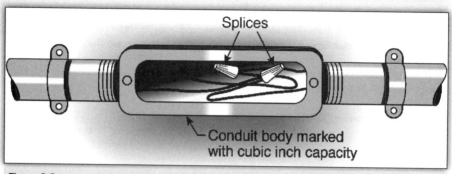

Splices

Conduit body marked
with cubic inch capacity

Figure 3.6 Conductor splices must be in boxes, conduit bodies, or fittings listed for the purpose. Conductors cannot be spliced within raceways. Reprinted with permission from 2005 NEC Handbook, copyright © 2005, National Fire Protection Association.

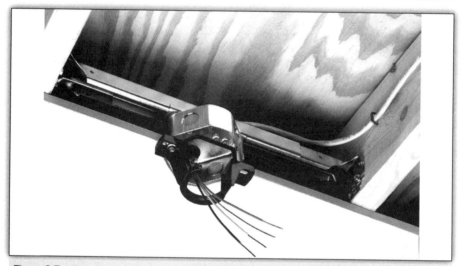

Figure 3.7 This fan-rated box has an adjustable bracket for positioning the unit between ceiling joists.

Electrical continuity of metal raceways and enclosures [300.10]. All metal parts of wiring systems must be bonded to form an effective low-impedance path to ground.

Firestopping [300.21]. All electrical penetrations through fire-rated walls, ceilings, and floors must be firestopped to maintain the fire-resistance rating. The *NEC* doesn't specify firestopping methods; these are covered in building codes such as NFPA 5000, *Building Construction and Safety Code* [NFPA 5000, 8.8.2].

Induced currents and heating [300.22, 300.35]. When alternating-current circuits are installed in metal raceways and enclosures, all circuit conductors must be grouped together to minimize unwanted induction.

Outlet boxes [314.27]. Special listed outlet boxes are required to support heavy luminaires [314.27(B)] and ceiling-suspended (paddle) fans [314.27(C)] (**Figure 3.7**).

Protection against physical damage [300.4]. When various types of cables and raceways are installed in holes bored through wood-framing members, nail protectors must be installed unless the cables or raceways can be kept at least 32 mm (1-¼ in) from the face of the wood-framing member. When nonmetallic-sheathed cable (Type NM) is installed through openings in metal-framing members, listed bushings or grommets must be installed to protect the cable from abrasion.

Securing and supporting [300.11]. Raceways, cables, boxes, and fittings must be securely fastened in place and adequately supported; they cannot be supported by ceiling grids. Raceways can sometimes support other raceways, cables, or equipment as described in 300.11(B), but cables are never permitted to support other items.

Underground cables and raceways. Burial depths are specified in Table 300.5.

Unused openings. Unused openings in enclosures must be adequately closed using metallic plugs or other means [110.12(A), 314.17(A)].

Wet locations. Metallic surface-mounted enclosures must be mounted with a minimum 6-mm (¼-in.) airspace between the box and the wall to prevent the entry of water [312.2(A)]; listed enclosures are manufactured with built-in "dimples" to meet this requirement. Cables and insulated conductors installed in underground raceways or enclosures must be listed for use in wet locations [300.5(B)]. Generally, but not always, these are conductors and cables with a "W" in their designation.

Cable and Raceway Articles

Most Chapter 3 articles are divided into three parts, with cable articles using a common numbering scheme and raceway articles using a common numbering scheme. Thus, the exposed work rules for armored cable (AC) are in 320.15 and the exposed work rules for non-metallic-sheathed cable (NM) are in 334.15. The listing requirement for intermediate metal conduit (IMC) is in 342.6 and the listing requirement for rigid nonmetallic conduit (RNC) is in 352.6.

This parallel numbering makes it easier to find similar information for different Chapter 3 wiring methods. The master numbering plans for cable and raceway articles are shown here (not every cable or raceway article contains every section shown):

Master Numbering Plans for Cable and Raceway Articles

Cable Articles Numbering Plan (Category 2)

Part I.   General

3xx.1   Scope

3xx.2   Definitions

3xx.6   Listing Requirements

Part II.   Installation

3xx.10   Uses Permitted

3xx.12   Uses Not Permitted

3xx.14   Dissimilar Metals

3xx.15   Exposed Work

3xx.16   Temperature Limits

3xx.17   Through or Parallel to Framing Members

3xx.18   Crossings

3xx.22   Number of Conductors

3xx.23   Inaccessible Attics

3xx.24   Bending Radius

3xx.30   Securing and Supporting

3xx.31   Single Conductors

3xx.40   Boxes and Fittings

3xx.41   Floor Coverings

3xx.42   Devices

3xx.56   Splices and Taps

3xx.60   Grounding and Bonding

3xx.80   Ampacity

Part III.   Construction Specifications

3xx.100   Construction

3xx.104   Conductors

3xx.108   Equipment Grounding

3xx.112   Insulation

3xx.116   Sheath, Jacket, or Conduit

3xx.120   Marking

Raceway Articles Numbering Plan (Categories 3 and 4)

Part I.   General

3xx.1   Scope

3xx.2   Definitions

3xx.3   Other Articles

3xx.6   Listing Requirements

Part II.   Installation

3xx.10   Uses Permitted

3xx.12   Uses Not Permitted

3xx.14   Dissimilar Metals

3xx.16   Temperature Limits

3xx.20   Size

3xx.22   Number of Conductors

3xx.24   Bends—How Made

3xx.26   Bends—Number in One Run

3xx.28   Reaming and Threading (or "Trimming" for nonmetallic raceways)

Understanding "Uses Permitted" and "Uses Not Permitted"

As discussed in Unit 1 of this book, the *National Electrical Code* is a permissive document. If a practice isn't specifically prohibited by the *NEC*, it is permitted. This means the lists of 3xx.10 "Uses Permitted" in most Chapter 3 wiring method articles are actually lists of typical examples. These wiring methods can legitimately be installed anywhere that isn't specifically *prohibited* in 3xx.12, "Uses Not Permitted."

However, the permissive nature of the *Code* isn't clear to many users—and over the years, these "Uses Permitted" lists turned out to be a confusing aspect of the *National Electrical Code*. Many professionals, including engineers, contractors, electricians, and inspectors, believed that the "Uses Permitted" lists were not typical examples, but absolute rules— that Chapter 3 wiring methods could be installed *only* as specified under "Uses Permitted."

For this reason, the 2005 *NEC* added this short FPN to the sections listed here:

FPN: The "Uses Permitted is not an all-inclusive list."
- Armored Cable: Type AC [320.10]
- Medium-Voltage Cable: Type MV [328.10]
- Metal-Clad Cable: Type MC [330.10]

Determining Raceway Sizes

Determining raceway sizes is simpler than calculating conductor sizes. The actual *Code* requirement is the very simple Table 1 of Chapter 9 (**Table 3.7**). The three specified conditions of permissible conduit fill aren't based on conductor heating considerations. These are dealt with when selecting conductor sizes, using the temperature correction factors and adjustment factors in Article 310. Instead, the two FPNs explain that the conduit fill percentages in Table 1 are intended to facilitate pulling conductors into raceways.

Table 3.7 Percent of Cross Section of Conduit and Tubing for Conductors

Number of Conductors	All Conductor Types
1	53
2	31
Over 2	40

FPN No. 1: Table is based on common conditions of proper cabling and alignment of conductors where the length of the pull and the number of bends are within reasonable limits. It should be recognized that, for certain conditions, a larger size conduit or a lesser conduit fill should be considered.

Annex C Tables

When all conductors in a conduit or tubing or circular cross-section are of the same size and insulation type, the tables in Annex C are the easiest and most practical way to determine raceway sizes. *NEC* users rely on these widely used tables as though they are part of the mandatory requirements of the *Code*.

Annex C has 12 pairs of tables, each covering one type of metal or nonmetallic raceway for both conventional and compact-stranded conductors. Confusingly, some of the table titles don't match the Chapter 3 names of these raceways; however, engineers soon become accustomed to using them. The Annex C conduit and tubing fill tables list raceways by both conventional trade sizes (nominal inch-pound designations) and equivalent metric designators.

The following excerpts from the *Code* illustrate the use of Annex C tables to select raceways for different installation situations.

Parking lot lighting.

An exterior lighting installation requires ten 10 AWG THWN-2 copper conductors in a raceway under a parking lot. What size PVC conduit is required?

Solution
There are four different types of PVC raceways in the *National Electrical Code*. Each is covered by a different conductor fill table in Annex C. The required raceway sizes are:

Table 3.8 Types of PVC Conduit

Wiring Method	Table	Minimum Trade Size Conduit
PVC Schedule 80	Table C.9	1
PVC Schedule 40 and HDPE	Table C.10	1
PVC Type A	Table C.11	¾
PVC Type EB	Table C.12	Available only in trade sizes 2 and larger

Motor circuit.

A 3-phase, 480 volt, 40hp motor is supplied by three 4 AWG THW conductors. What size metal conduit or tubing will be required? What size flexible metal conduit will be required at the motor termination?

Solution

Annex C tables for various types of metal conduit and tubing, flexible metal conduit, and Liquidtight flexible metal conduit give the minimum sizes shown below:

Table 3.9 Types of Metal Raceways

Wiring Method	Table	Minimum Trade Size Conduit
EMT	Table C.1	1
IMC	Table C.4	1
RMC	Table C.8	1
FMC	Table C.3	1
Liquidtight FMC	Table C.7	1

Underground service.

An underground service lateral requires four 600 kcmil XHHW compact-stranded aluminum conductors. What trade size conduit is required for rigid metallic conduit (Type RMC), intermediate metal conduit (IMC), PVC Schedule 40, PVC Schedule 80, and PVC Type EB?

Solution

Annex C tables list the minimum trade sizes for each conduit type needed for the four 600 kcmil aluminum conductors. Compact-stranded aluminum building wire in Types THW, THWN/THHN, and XHHW is compact stranded are covered in the "A" table for each of the five conduit types.

Table 3.10

Wiring Method	Table	Minimum Trade Size Conduit
RMC	Table C.8A	3
IMC	Table C.4A	3
PVC Schedule 40	Table C.10A	3
PVC Schedule 80	Table C.9A	3 ½
PVC Type EB	Table C.12A	3

Conclusion

This unit discusses conductors, circuits, and wiring methods. The purpose of the *National Electrical Code* is the "practical safeguarding of persons and property from hazards rising from the use of electricity," and protection of conductors is one of its primary concerns. Safety rules governing the proper use of conductors and raceways are located throughout the *Code*, but are concentrated in *NEC* Chapters 2 and 3. This is the first of four units that explains how actual *NEC* rules affect the way in which engineers make design decisions and choose among competing methods and technologies.

UNIT 4

Electrical Equipment

Introduction

The *National Electrical Code* is a complex and comprehensive document that attempts to organize a broad range of equipment and systems into a logical framework. This structure is useful and necessary, but for new *Code* users, it sometimes obscures the interrelated and interconnected nature of *NEC* requirements for electrical products and systems. This unit provides an overview of *Code* safety rules primarily related to all types of electrical equipment. They can be summarized as follows:

- Chapter 1 covers general conditions for all electrical installations, including many related to electrical equipment.
- Chapter 2 contains many requirements related to equipment, including load calculations, services, overcurrent protection, grounding and bonding, and two articles on different types of equipment that protect against surge protection and transients.
- Chapter 3 contains two articles on boxes, enclosures, and equipment cabinets.
- Chapter 4 covers equipment for general use. Products covered include equipment for generating, conditioning, and distributing power; utilization equipment including luminaires; motors and their associated controls; HVAC (heating, ventilating, and air conditioning) equipment; electric appliances; and industrial heating equipment.
- Chapter 5 on special occupancies contains many requirements governing the construction and installation of electrical equipment. These include equipment used in hazardous (classified) locations; temporary power systems and equipment used for building construction, carnivals, fairs, trade shows, and similar events; electrical equipment of vehicles, mobile homes, and manufactured buildings; and areas characterized by large concentrations of people.
- Chapter 6 deals with a broad range of specialized equipment including industrial machinery, information technology equipment, electric signs and outdoor lighting, cranes and hoists, welders, swimming pools and fountains, and photovoltaic systems. Although Chapter 7 typically is considered the "emergency systems chapter" of the *Code*, fire pumps are included at the end of Chapter 6.
- Chapters 7 and 8 cover special systems and communications systems, and include rules related to conductors, cables, raceways, and equipment. Because most of these systems are low voltage and limited energy, they aren't covered by the general rules of *NEC* Chapters 1–4. These are covered in Unit 5 of this book
- Chapter 9 does not contain information related to equipment.

Chapter 1—General

Article 100 definitions were covered in Unit 1 of this book. Key rules of Article 110 are summarized here, along with related general requirements for installing electrical equipment that appear in other articles of the *Code*.

⚠WARNING

Potential Arc Flash Hazard
Appropriate PPE and Tools Required when working on this equipment

Figure 4.1   *NEC* 110.16 requires arc-flash warning signs on most electrical distribution equipment. Reprinted with permission from 2005 NEC Handbook, copyright © 2005, National Fire Protection Association.

Interrupting Rating

Equipment must have an interrupting rating sufficient to withstand the available fault current and voltage [100.9]. Current-limiting overcurrent protection devices [240.2] are often installed in service equipment to meet this requirement.

Flash Protection

Most types of power distribution equipment and meter socket enclosures must be marked to warn qualified persons of potential arc flash hazards that require the use of appropriate personal protective equipment (PPE) when working on energized conductors and equipment (**Figure 4.1**). Although 110.16 does not require detailed marking of the available incident energy on the equipment, this practice is becoming common in many industrial occupancies.

Spaces Around Electrical Equipment

The *NEC* has detailed rules specifying adequate space around electrical equipment, so that the equipment can be serviced safely while energized, should this be necessary. Key concepts include the following:

Working Space [110.26(A) and (B)]

This term generally applies to protection of workers who may need to examine or service electrical equipment under energized conditions. Working space cannot be used for storage. Working space also is required around electric equipment located outdoors. Although

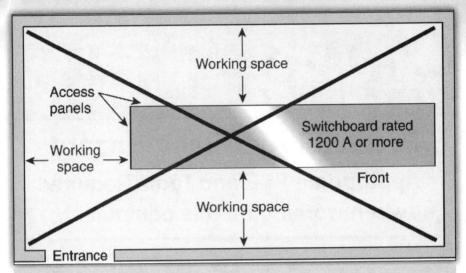

Figure 4.2 This arrangement of a large switchboard is unacceptable per 110.26(C). A person could be trapped behind arcing electrical equipment. Reprinted with permission from 2005 NEC Handbook, copyright © 2005, National Fire Protection Association.

confusingly, this rule is located in 110.26(F)(2), which otherwise concerns *dedicated equipment space* (explained later).

Entrance to working space. The number and locations of required entrances to electric equipment rooms depends on the ampere rating and configuration of the equipment within [110.26(C)]. The intent is that workers shall be able to escape from an electrical equipment room, in an emergency, without being trapped behind arcing electrical equipment (**Figure 4.2**).

For power distribution and control equipment rated 1,200 amperes or more, 110.26(C) requires that personnel doors open in the direction of egress and be equipped with panic bars or other devices that open under simple pressure. This is one of very few places that the *NEC* includes information normally contained in nonelectrical building codes. Also see the section on transformer vaults, later in this unit.

Dedicated Equipment Space

This term generally refers to space reserved for installation of future electrical equipment and wiring methods, and to protecting the equipment from intrusion by nonelectrical equipment. *NEC* 110.26(F)(1)(c) specifically permits sprinklers in electrical equipment rooms.

Headroom and Mounting Heights of Equipment

Headroom. The minimum headroom of working space about service equipment, switchboards, panelboards, and motor control centers is 2.0 m (6 ½ ft) [110.26(E)]. In many ways, motor control centers are similar in construction to switchboards and are permitted to serve as service equipment. In existing dwellings, service equipment not exceeding 200 amperes can be installed in spaces with lower headroom.

Mounting heights. Switches and circuit breakers used as switches must be located so they are readily accessible, and so that the operating handle is a maximum of 2.0 m above the floor or working platform in its highest position [404.8(A)]. The *Code* also contains other rules that specify *minimum* mounting heights for various types of equipment:

- *Fire Pumps.* All energized parts of fire pump installations must be located at least 300 mm (12 in.) above floor level [695.12(D)].
- *Mobile Homes.* Outdoor disconnecting means must be installed not less than 600 mm (2.0 ft) above the grade or working platform [550.32(F)].
- *Recreational Vehicles.* Site supply equipment must be installed not less than 600 mm (2.0 ft) above grade.
- *Marinas and Boatyards.* Receptacles at marinas and boatyards shall be located at least 300 mm (12 in.) above decks of piers [555.19].

Illumination

The *NEC* doesn't specify illumination levels for electrical equipment rooms located indoors. It also doesn't require artificial lighting for electrical equipment located outdoors. The last sentence of 110.26(D) means that illumination in electrical equipment rooms cannot be controlled by timer switches, occupancy sensors, or central lighting control systems unless there is a manual override located in the room. The intent of this requirement is to ensure that lighting is not extinguished unexpectedly while personnel are working around energized equipment.

Openings Must Be Closed

Unused openings in raceways, enclosures, and equipment cases must be closed to prevent accidental contact with energized conductors inside [110.12, 314.7, 408.7].

Chapter 2—Wiring and Protection

Despite the name of this *NEC* chapter, many of its rules concern equipment. These are summarized here, along with related requirements for installing electrical equipment that appear in other articles of the *Code*.

Load Calculations

Although Article 220 covers branch-circuit, feeder, and service calculations, its requirements are based on loads of electric equipment. Table 220.3 contains nearly 30 references to additional load calculation references, involving many different types of electrical equipment (**Table 4.1**).

Table 4.1 Additional Load Calculation References

Calculation	Article	Section (or Part)
Air-Conditioning and Refrigerating Equipment, Branch-Circuit Conductor Sizing	440	Part IV
Cranes and Hoists, Rating and Size of Conductors	610	610.14
Electric Welders, Ampacity Calculations	630	630.11, 630.31
Electrically Driven or Controlled Irrigation Machines	675	675.7(A), 675.22(A)
Electrolytic Cell Lines	668	668.3(C)
Electroplating, Branch-Circuit Conductor Sizing	669	669.5
Elevator Feeder Demand Factors	620	620.14
Fire Pumps, Voltage Drop (mandatory calculation)	695	695.7
Fixed Electric Heating Equipment for Pipelines and Vessels, Branch-Circuit Sizing	427	427.4
Fixed Electric Space Heating Equipment, Branch-Circuit Sizing	424	424.3
Fixed Outdoor Electric Deicing and Snow-Melting Equipment, Branch-Circuit Sizing	426	426.4
Industrial Machinery, Supply Conductor Sizing	670	670.4(A)
Marinas and Boatyards, Feeder and Service Load Calculations	555	555.12
Mobile Homes, Manufactured Homes, and Mobile Home Parks, Total Load for Determining Power Supply	550	550.18(B)
Mobile Homes, Manufactured Homes, and Mobile Home Parks, Allowable Demand Factors for Park Electrical Wiring Systems	550	550.31
Motion Picture Television Studios and Similar Locations— Sizing of Feeder Conductors for Television Studio Sets	530	530.19
Motors, Feeder Demand Factor	430	430.26
Motors, Multimotor and Combination-Load Equipment	430	430.25
Motors, Several Motors or a Motor(s) and Other Load(s)	430	430.24
Over 600 Volt Branch-Circuit Calculations	210	210.19(B)
Over 600 Volt Feeder Calculations	215	215.2(B)
Phase Converters, Conductors	455	455.6
Recreational Vehicle Parks, Basis of Calculations	551	551.73(A)
Sensitive Electrical Equipment, Voltage Drop (mandatory calculation)	647	647.4(D)
Solar Photovoltaic Systems, Circuit Sizing and Current	690	690.8
Storage-Type Water Heaters	422	422.11(E)
Theaters, Stage Switchboard Feeders	520	520.27

Reprinted with permission from 2005 NEC Handbook, copyright © 2005, National Fire Protection Association.

Services

The *National Electrical Code* requires that each building or other structure be supplied by only one service, feeder, or branch circuit, except under special circumstances [225.30, 230.2]. It also requires that:

- Each service or other supply shall have a maximum of six disconnects, either switches or circuit breakers [225.33(A), 230.71(A)].
- Fuses by themselves are not permitted to serve as disconnects, because they cannot simultaneously disconnect all phase (ungrounded) conductors. However, fused disconnect switches are commonly installed [225.38(B), 230.74].
- The (up to six) disconnects shall be grouped [225.34(A), 230.72(A)].
- The disconnects can be installed either indoors or outdoors. They must be located at a readily accessible location nearest the point where the service, feeder, or branch circuit conductors enter the building or structure [225.32, 230.70(A)].

Safety Rationale

These *NEC* rules for services and disconnects are intended to make it easy for fire fighters and other emergency responders to locate and turn off the electric supply to a building or other structure. They also minimize the length of service conductors inside a building or other structure that don't have overcurrent protection to *NEC* rules.

Conductors considered outside buildings. Conductors are considered to be outside a building or other structure when installed beneath the structure under at least 50 mm (2 in.), or several other conditions defined in 230.6.

Overcurrent Protection

Article 240 provides general requirements for overcurrent protection of both conductors and equipment. Section 240.3 provides nearly 40 references to other articles for more specialized overcurrent protection rules for specific types of equipment and occupancies.

Coordinated (Selective) Overcurrent Protection

In situations where an orderly shutdown is required, coordinated overcurrent protection is permitted to control and limit the extent of power losses (**Figure 4.3**). Article 620 also contains detailed requirements on coordinated overcurrent protection, because elevators and similar equipment are frequently used by firefighter and rescue personnel during emergencies. Thus, if a building contains three elevators and a fault occurs in one, the overcurrent device ahead of the faulted elevator should open, leaving the other two in operation.

Fuses and Circuit Breakers

The *NEC* doesn't have separate articles devoted to either type of overcurrent protection device.

- Fuses are covered in Parts V and VI of Article 240.
- Circuit breakers are covered in Part VII of Article 240 (**Figure 4.4**).

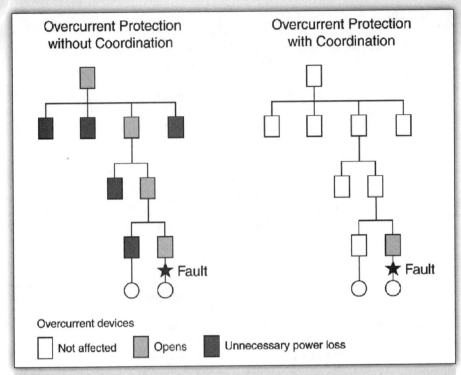

Figure 4.3 Overcurrent protection schemes can be either with coordination or without.

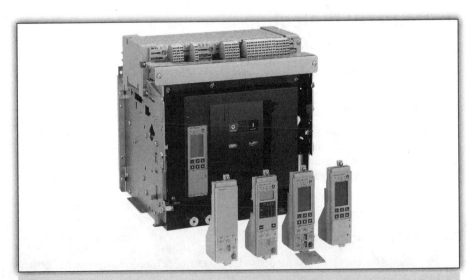

Figure 4.4 This adjustable trip circuit breaker has a transparent, removable, sealable cover.

Interrupting ratings. Fuses have an interrupting rating of 10,000 amperes unless otherwise marked [240.60(C)(3)]. Circuit breakers have interrupting ratings of 5,000 amperes unless otherwise marked [240.83(B)].

Circuit breakers used as switches. Circuit breakers marked "SWD" can be used as switching devices for fluorescent lighting loads. Circuit breakers marked "HID" can be used for switching both fluorescent and high-intensity discharge lighting loads [240.83(D)].

Grounding and Bonding

Article 250 covers requirements for grounding and bonding of electrical installations. It also specifies the conditions under which guards, isolation, or insulation may be substituted for grounding. Article 250 is one of the longest articles in the *Code*, and many users find it one of the most complex to understand and apply. In recognition of this fact, Figure 250.1 in the *NEC* provides a kind of "road map" that shows how the various parts of the article relate to one another (**Figure 4.5**).

Grounding versus bonding. Although grounding and bonding are related concepts, Article 100 defines them differently:

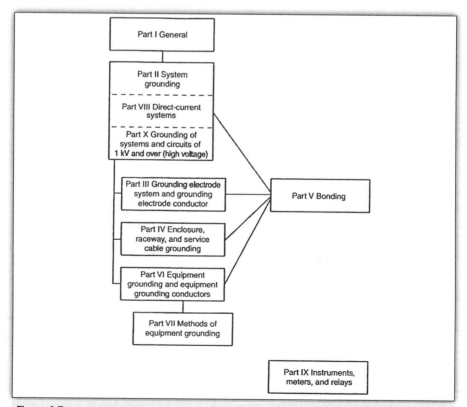

Figure 4.5 Grounding and bonding. Reprinted with permission from 2005 NEC Handbook, copyright © 2005, National Fire Protection Association.

Grounded (Grounding). Connected to ground or to a conductive body that extends the ground connection.

Bonded (Bonding). Connected to establish electrical continuity and conductivity.

Thus, *grounded* and *grounding* refer to equipment and conductors connected to the earth, while *bonded* and *bonding* refer to equipment and conductors connected to each other to form a fault current path to the earth. The purpose of bonding is to establish an effective path for fault current that facilitates operation of overcurrent protective devices [250.4(A)(3) and (4), 250.4(B)(3) and (4)].

The terminology doesn't always make this distinction easy to understand; many *Code* experts feel that *equipment grounding conductors* should be called *equipment bonding conductors*.

Equipotential bonding. *NEC* 250.3 refers to more specialized grounding and bonding requirements located in other articles of the *Code*. One that shows up in three different articles is equipotential bonding:

- Article 680—Swimming Pools, Fountains, and Similar Installations
- Article 682—Natural and Artificially Made Bodies of Water
- Article 547—Agricultural Buildings

These three articles require that a common bonding grid be established in areas where water is used to create equal electrical potential (voltage) that minimizes voltage gradients, thereby minimizing touch and step voltage. In the case of swimming pools and electrical equipment of other artificial bodies of water (**Figure 4.6**), the purpose is personnel protection. In the case of agricultural buildings, equipotential grounding is productivity related, because animals such as dairy cattle are sensitive to stray currents.

Interestingly, the term *equipotential plane* is defined in 547.2, though it is also used in 682.33. The term doesn't appear in Article 680, which refers instead to *equipotential bonding* in 680.26.

Surge and Transient Protection

Article 280—Surge Arresters covers requirements for surge arresters that protect people and equipment from high-voltage surges caused by lightning or an electrical fault in the utility network. Surge arresters (also known as *lightning arresters*) are typically installed on the load side of the service equipment, though this isn't the only point at which they can be used.

Article 285—Transient Voltage Surge Suppressors (TVSSs) covers several different types of listed surge suppressors installed on premise wiring systems, including receptacles with integral TVSSs, computer power strips with built-in metal-oxide varistors (MOV), and units installed at panelboard.

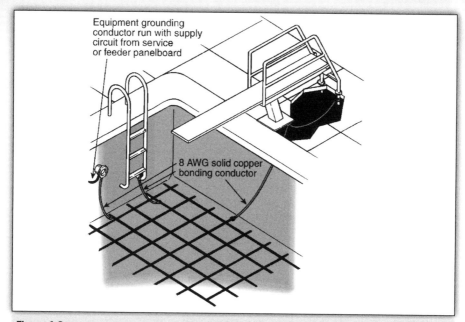

Equipment grounding
conductor run with supply
circuit from service
or feeder panelboard

8 AWG solid copper
bonding conductor

Figure 4.6 Equipotential bonding is used to reduce shock hazards at swimming pools. Reprinted with permission from 2005 NEC Handbook, copyright © 2005, National Fire Protection Association.

Chapter 3—Wiring Materials and Methods

Chapter 3 contains two articles covering enclosures. Although most engineers and other electrical professionals regard enclosures primarily as metal or nonmetallic boxes, *NEC* Article 100 defines the term more broadly:

Enclosure. The case or housing of apparatus, or the fence or walls surrounding an installation to prevent personnel from accidentally contacting energized parts or to protect the equipment from physical damage.

FPN: See Table 110.20 for examples of enclosure types.

Thus, Article 100 defines *enclosure* in two very different ways:

1. As a box that prevents persons from contacting energized parts inside
2. As a fence or wall that prevents damage to the electrical equipment itself

The FPN reinforces the confusion by referring to an environmental classification system for metal and nonmetallic boxes. Part V of Article 110 adds another dimension by defining rules for "manholes and other enclosures intended for personnel entry."

The *NEC* Chapter 3 rules for enclosures deal with metal and nonmetallic boxes designed to contain electric conductors and equipment.

Article 312—Cabinets, Cutout Boxes, and Meter Socket Enclosures

This article contains construction specifications, and general rules such as minimum wire bending space, for enclosures designed primarily to contain electrical equipment such as panels and disconnect switches. Although basic enclosure rules appear in Chapter 3, most types of enclosed equipment are covered in Chapter 4.

Article 314—Outlet, Device, Pull and Junction Boxes; Conduit Bodies; Fittings; and Handhole Enclosures

This article covers enclosures designed primarily to enclose conductors and wiring devices such as wall switches, receptacles, and dimmers. It includes construction specifications and general rules such as conductor fill requirements and rules for pulling conductors through enclosures (i.e., straight pulls, angle pulls, and U-pulls).

Enclosure Types

Electrical enclosures are listed for protection against environmental conditions according to a classification scheme based on NEMA and UL standards. These types are described in Table 430.91 in the *NEC* article of motors and controllers, although they apply to all electrical enclosures. The most commonly used North American electrical enclosure types are:

- Type 1—General (unmarked enclosures are Type 1)
- Type 3R—Rainproof
- Type 4—Watertight
- Type 12—Dusttight

Ingress protection. The *National Electrical Code* doesn't use IP enclosure types defined by IEC standards.

Chapter 4—Equipment for General Use

Chapter 4 is the primary equipment chapter of the *NEC*. Equipment covered by its rules is installed in most kinds of commercial, institutional, and residential buildings. It is divided into 21 articles covering major subjects:

- Article 400—Flexible Cords and Cables
- Article 402—Fixture Wires
- Article 404—Switches
- Article 406—Receptacles, Cord Connectors, and Attachment Plugs (Caps)
- Article 408—Switchboards and Panelboards
- Article 409—Industrial Control Panels
- Article 410—Luminaires, Lampholders, and Lamps
- Article 411—Lighting Systems Operating at 30 Volts or Less
- Article 422—Appliances
- Article 424—Fixed Electric Space-Heating Equipment

- Article 426—Fixed Outdoor Electric Deicing and Snow-Melting Equipment
- Article 427—Fixed Electric Heating Equipment for Pipelines and Vessels
- Article 430—Motors, Motor Circuits, and Controllers
- Article 440—Air-Conditioning and Refrigerating Equipment
- Article 445—Generators
- Article 450—Transformers and Transformer Vaults (Including Secondary Ties)
- Article 455—Phase Converters
- Article 460—Capacitors
- Article 470—Resistors and Reactors
- Article 480—Storage Batteries
- Article 490—Equipment, Over 600 Volts, Nominal

Key Concepts

The 21 articles of Chapter 4 contain both detailed requirements for installing electrical equipment, and safety-related design and performance requirements that serve as a guide for manufacturers and a basis for product listing. The range of equipment covered by Chapter 4 is too wide to summarize in a short space. Important concepts that engineers designing and specifying electrical installations must understand include the following.

Disconnecting Means

All *Code* articles dealing with equipment have requirements for disconnecting means that permit equipment to be disconnected from its power supply for servicing:

- Disconnecting means must simultaneously interrupt all grounded (phase) conductors, except under special circumstances. Grounded (neutral) conductors aren't normally switched or disconnected under *NEC* rules.
- Branch-circuit overcurrent devices and or an attachment plug and receptacle are permitted to serve as disconnecting means for utilization equipment, under certain circumstances [422.33, 440.63].
- On-off switches built into appliances generally aren't permitted to be the only disconnecting means.
- Normally, disconnecting means are required to be within sight of the equipment controlled, unless it is capable of being locked in the open position. This allows the qualified person doing the servicing to verify that the disconnecting means is off and the utilization equipment is deenergized. Article 100 defines this as follows:

In Sight From (Within Sight From, Within Sight). Where this *Code* specifies that one equipment shall be "in sight from," "within sight from," or "within sight," and so forth, of another equipment, the specified equipment is to be visible and not more than 15 m (50 ft) distant from the other.

Flexible Cords and Cables; Fixture Wires

As explained in Unit 3 of this book, flexible cords and cables (Article 400) and fixture wires (Article 402) aren't considered wiring methods, and can't be used as a substitute for fixed wiring or where concealed behind walls, floors, or ceilings [400.7, 402.11].

Flexible cords and cables typically are used where flexibility is needed due to movement of equipment, frequent repositioning, or (like flexible raceways) to provide vibration isolation at connections to motorized equipment. Flexible cords and cables are extensively used with the following types or equipment:

- Appliances (Article 422)
- Audio systems (Article 640)
- Carnivals, circuses, fairs, and similar events (Article 525)
- Cranes and hoists (Article 610)
- Electrical vehicle charging systems (Article 625)
- Elevators and similar equipment (Article 620)
- Information technology equipment (Article 645)
- Luminaires and lighting systems (Articles 410, 411)
- Marinas, boatyards, and floating buildings (Articles 553, 555)
- Mining and excavating equipment (Article 490)
- Mobile homes, recreational vehicles, and park trailers (Articles 550, 551, 552)
- Motion picture and TV studios and theaters (Article 530)

Lighting Equipment

Article 410—Luminaires, Lampholders, and Lamps provides the main rules for luminaires and associated equipment, including neon and cold-cathode lighting, located both indoors and outdoors. Key concepts include the following:

Clothes Closets

To keep hot lamps away from combustible clothing, luminaires are not permitted within the storage space illustrated by *NEC* Figure 410.8 (**Figure 4.7**). Incandescent fixtures with exposed lamps, and pendant fixtures or lampholders, aren't permitted within clothes closets at all.

Luminaires Supported by Trees

This is specifically permitted by 410.16(H). However, the *Code* doesn't permit overhead conductor spans to be supported by vegetation [225.26, 230.10].

Luminaires Used as Raceways

Luminaires used as raceways for through-wiring of branch-circuits to other fixtures must be listed and marked for that purpose [410.31]. This is most often done with installations of fluorescent fixtures mounted end-to-end.

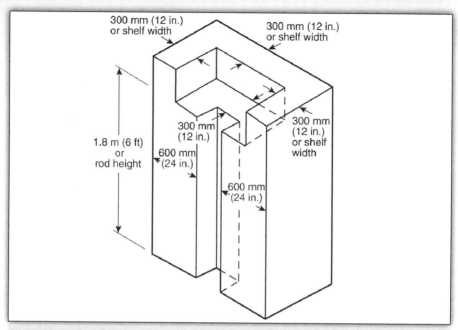

Figure 4.7 Closet storage space. Reprinted with permission from 2005 NEC Handbook, copyright © 2005, National Fire Protection Association.

Lighting Poles Used as Raceways

Metal poles for outdoor lighting fixtures aren't permitted to contain low-voltage or limited-energy circuits. Specifically, 410.15(B) defines them as a "raceway to enclose supply conductors." The intent is to prevent conductors for security cameras, loudspeakers, and so on from being installed inside lighting poles with power wiring. The *NEC* usually tries to keep power conductors separate from low-voltage signaling and communications wiring.

Thermal Considerations

Thermal protectors. Recessed incandescent and high-intensity discharge (HID) luminaires are required to have an integral thermal protector that opens automatically in case of overheating to cut off electricity to the lamp [410.65].

Class P ballasts. Most fluorescent luminaires installed indoors have listed Class P ballasts that automatically disconnect the power supply when their case temperature exceeds 90°C (194°F) [410.73(E)(1)]. However, fluorescent fixtures approved for use as emergency lighting don't have Class P ballasts, so they will continue to operate under high-temperature conditions. This requirement isn't a *Code* rule *per se*, but appears in the UL listing standard for emergency-type fluorescent fixtures.

Thermal insulation. Recessed incandescent and HID luminaires installed in direct contact with flammable thermal insulation (**Figure 4.8**) are required to be listed and identified as Type IC [410.66(A)(2)]. These are most often used in one- and two-family dwellings.

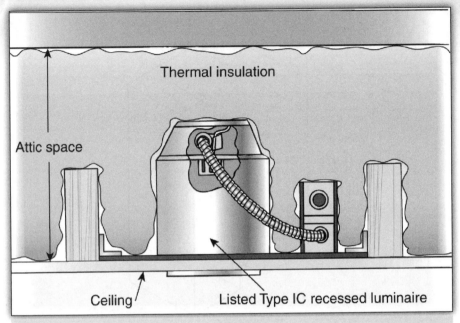

Figure 4.8 Recessed luminaires identified as Type IC are constructed so that the outsides of their cans (or enclosures) don't reach high temperatures and can safely be installed in direct contact with thermal insulation.

The general Article 410 rules for lighting equipment are modified for more specialized applications by:

- Article 411—Lighting Systems operating at 30 Volts or Less
- Article 600—Electric Signs and Outline Lighting

Motors and Motor-Driven Equipment

Article 430 covers motors, branch circuits, and feeders for motors, motor overload protection, motor controllers and control circuits, and motor control centers. It is one of the two longest, most complex articles in the *National Electrical* Code (the other being Article 250 on grounding and bonding). Article 430 is divided into 14 parts:

   I. General

  II. Motor Circuit Conductors

 III. Motor and Branch-Circuit Overload Protection

 IV. Motor Branch-Circuit Short-Circuit and Ground-Fault Protection

  V. Motor Feeder Short-Circuit and Ground-Fault Protection

 VI. Motor Control Circuits

VII. Motor Controllers

There are four major reasons for the complexity of Article 430:

1. *Motor designs.* Electric motors are manufactured in many different types, with different operating characteristics for different applications.

2. *High motor starting currents.* Most types of motors have starting inrush currents much higher than their full-load operating currents. This problem complicates motor overload and overcurrent protection: Devices must be set at ampere levels high enough to permit motors to start but low enough to limit steady-state operating current to safe levels within the motor's design characteristics.

3. *Motor control technology.* Largely because of these high inrush currents, there are many different types of motor controllers used with different types and sizes of motors. A fractional horsepower motor may be controlled by a simple snap switch, a small integral horsepower motor by an across-the-line motor starter, and a large motor by a part-winding motor starter. Adjustable-speed drive systems introduce an additional degree of complexity.

4. *Horsepower ratings.* Although most electrical equipment is rated in volt-amperes (VA), motors have traditionally been rated in horsepower (hp). Tables 430.247 through 430.251(B) provide conversions of motor horsepower ratings into amperes for full-load currents and locked-rotor currents. Some motors have their output ratings expressed in watts and kilowatts (1 hp equals approximately 746 watts). However, it is important to follow 430.6 for determining ampacity and motor rating; sizing circuits based solely on kilowatt output can result in undersized conductors and overcurrent protective devices.

Key concepts that engineers must understand to apply Article 430 include the following:

Current ratings. These are generally determined according to the values in Tables 430.247 through 430.251 (B).

Design letters. These are shown on motor nameplates and refer to motor construction and operating characteristics; Design B is the most common motor type.

Locked-rotor-indicating code letters. Locked-rotor-indicating letters are used to determine the maximum locked-rotor current for a particular type of motor. These code letters also are shown on motor nameplates and must not be confused with *design letters.*

Protection. In addition to branch-circuit overcurrent protection required by Article 240, motors must have thermal overload protectors (often called *heaters* in the field), motor branch circuits must have short-circuit and ground-fault protection, and feeders serving motor loads are required to have short-circuit and ground-fault protection.

Air-Conditioning and Refrigerating Equipment

Article 440 supplements the general requirements of Article 430 for motor-driven air-conditioning and refrigerating equipment that uses hermetic refrigerant motor-compressors. Because the motors for this type of equipment operate within coolant, they have different load and heating characteristics than other types of motors.

Panelboards

General requirements for panelboards include the following. The commonly-used term *main distribution panel* doesn't appear in the *NEC*.

408.4 Circuit Directory or Circuit Identification. Every circuit and circuit modification shall be legibly identified as to its clear, evident, and specific purpose or use. The identification shall include sufficient detail to allow each circuit to be distinguished from all others. Spare positions that contain unused overcurrent devices or switches shall be described accordingly. The identification shall be included in a circuit directory that is located on the face or inside of the panel door in the case of a panelboard, and located each switch on a swtichboard. No circuit shall be described in a manner that depends on transient conditions of occupancy.

Overcurrent Protection

The main overcurrent device for a panelboard is normally within the panelboard itself or on its supply side (such as a subfeed circuit breaker or fused switch in a switchboard or other panelboard). However, 408.36(B) permits a panelboard supplied through a transformer to be protected by the transformer's primary overcurrent device, under certain circumstances.

Maximum Number of Devices

A long-standing *Code* rule that limited panelboards to a maximum of 42 overcurrent devices (defined as pole spaces) was removed from the 2008 edition. This limit now applies only under very specialized circumstances described in 408.36, Exception No. 2. For general applications, the *NEC* no longer specifies a maximum number of overcurrent devices in a panelboard.

Service Equipment

Panelboards used as service equipment are not required to be protected by a single over-current device. Instead, they can have up to six main circuit breakers or fused switches, as permitted by 225.33 and 230.71(A).

Panelboards listed for use as service equipment also come supplied with a main bonding jumper (MBJ) that connects the grounded conductor terminal bar to the equipment grounding terminal bar and the panelboard enclosure [250.24(B), 250.28].

Load Center

This is a marketing term that doesn't appear in the *NEC* or the UL product standard for panelboards. Load centers are a type of panelboard used primarily in residential and light-commercial construction. They typically have cabinets 14 inches wide (for installation between studs 16 inches on center) and use circuit breakers ½ inch wide. By contrast, panelboards used in commercial-industrial-institutional construction typically have cabinets 20 inches wide and use circuit breakers 1 inch wide.

Storage Batteries and Electrolytic Cells

Article 480 covers battery installations used as a storage medium and source of electric energy in uninterruptible power supplies (UPS), emergency standby power systems, and photovoltaic installations. Electrolytic cells used for industrial process purposes are covered in Article 668.

Transformers

General-purpose transformers used in electric power systems to transform energy for purposes of efficient distribution, and to match voltage and current characteristics to those of the supplied utilization equipment, are covered by the rules of Article 450. Eight exceptions to the scope describe other types of specialty transformers covered elsewhere in the *NEC*.

Overcurrent Protection

The complex rules for overcurrent protection rules of transformers are presented in Table 450.3(A). Note 2 of this table allows overcurrent protection on the secondary side of transformers to consist of up to six sets of fuses or six circuit breakers, grouped in one location, an interesting parallel to the rules for service disconnects in 225.33(A) and 230.71(A).

Transformer Vaults

Because of the fire risk posed by flammable transformer insulating fluids, Article 450 contains detailed construction and fire-resistance requirements for transformer vaults.

450.42 Walls, Roofs, and Floors. The walls and roofs of vaults shall be constructed of materials that have adequate structural strength for the conditions with a minimum fire resistance of 3 hours. The floors of vaults in contact with the earth shall be of concrete that is not less than 100 mm (4 in.) thick, but where the vault is constructed with a vacant space or other stories below it, the floor shall have adequate structural strength for the load imposed thereon and a minimum fire resistance of 3 hours. For the purposes of this section, studs and wallboard construction shall not be acceptable. *Exception: Where transformers are protected with automatic sprinkler, water spray, carbon dioxide, or halon, construction of 1-hour rating shall be permitted.*

FPN No. 1: For additional information, see ANSI/ASTM E119-1995, *Method for Fire Tests of Building Construction and Materials*, and NFPA 251-2006, *Standard Methods of Tests of Fire Endurance of Building Construction and Materials*.

FPN No. 2: A typical 3-hour construction is 150 mm (6 in.) thick reinforced concrete.

Chapter 5—Special Occupancies

NEC Chapter 5 is divided into 28 articles covering major subjects:

- Article 500—Hazardous (Classified) Locations, Classes I, II, and III, Divisions 1 and 2
- Article 501—Class I Locations
- Article 502—Class II Locations
- Article 503—Class III Locations
- Article 504—Intrinsically Safe Systems
- Article 505—Class I, Zone 0, 1, and 2 Locations
- Article 506—Zone 20, 21, and 22 Locations for Combustible Dusts or Ignitible Fibers/Flyings
- Article 510—Hazardous (Classified) Locations—Specific
- Article 511—Commercial Garages, Repair and Storage
- Article 513—Aircraft Hangars
- Article 514—Motor Fuel Dispensing Facilities
- Article 515—Bulk Storage Plants
- Article 516—Spray Application, Dipping, and Coating Processes
- Article 517—Health Care Facilities
- Article 518—Assembly Occupancies
- Article 520—Theaters, Audience Areas of Motion Picture and Television Studios, Performance Areas, and Similar Locations
- Article 522—Control Systems for Permanent Amusement Attractions
- Article 525—Carnivals, Circuses, Fairs, and Similar Events
- Article 530—Motion Picture and Television Studios and Similar Locations
- Article 540—Motion Picture Projection Rooms
- Article 545—Manufactured Buildings
- Article 547—Agricultural Buildings
- Article 550—Mobile Homes, Manufactured Homes, and Mobile Home Parks
- Article 551—Recreational Vehicles and Recreational Vehicle Parks
- Article 552—Park Trailers
- Article 553—Floating Buildings
- Article 555—Marinas and Boatyards
- Article 590—Temporary Installations

Key Concepts

The *Code* articles grouped together as "special occupancies" contain a number of requirements for electrical equipment and systems based on where it is installed. Important concepts include the following:

Assembly occupancies. Assembly occupancies are those defined in NFPA 101, *Life Safety Code,* as being designed or intended for 100 or more persons. They include buildings such as churches, chapels, auditoriums, and restaurants. They don't include structures such as office buildings, schools, supermarkets, and hotels, even those that may contain 100 or more people, because these aren't specifically intended for the assembly of persons. However, auditoriums, cafeterias, restaurants, and similar areas within these building types are considered assembly occupancies. *NEC* requirements for assembly occupancies are specified in Article 518.

Fire-rated construction. Assembly occupancies are required by building codes such as NFPA 5000, *Building Construction and Safety Code,* to use fire-rated construction methods. *NEC* Article 518 requires that wiring methods used in fire-rated areas must be metal raceways, nonmetallic raceways encased in concrete or installed behind a thermal barrier with a minimum 15-minute finish rating, or Type MI, MC, or AC cable. Areas of nonrated construction within assembly occupancies, such as washrooms and offices, are permitted to use a wider range of wiring methods.

Theaters and Similar Locations

Theaters, audience areas of motion picture and television studios, and similar locations aren't covered by Article 518. They have their own Article 520 because of the large amount of special electrical equipment, including stage switchboards, dimming systems, portable power components, and extensive use of flexible cords and cables. Articles 530 and 540 contain electrical rules for other areas of motion picture and television studios.

Carnivals, Circuses, Fairs, and Similar Events

NEC Article 525 covers portable wiring and equipment used for both indoor and outdoor temporary events, including carnivals, circuses, fairs, craft shows, trade shows, wedding pavilions, and rock concerts. This article covers wiring on all structures such as carnival rides and lighting towers.

Other Articles

Audio systems are covered by Article 640. Water attractions such as fountains and pools are covered by the applicable requirements of Article 680. Temporary power systems used for construction, demolition, repair, and similar activities are covered by Article 590 (see the section on temporary installations, later in this unit.)

Major Concerns

Particular concerns of Article 525 include an emphasis on flexible cords and cables, use of separately derived power systems (generators and transformers), prevention of tripping hazards, and shock protection. Section 525.23 requires that most receptacles have ground-fault circuit-interrupter (GFCI) protection. Locking-type receptacles are exempted by 525.23(B).

Fire Pumps

Although most *Code* users regard Chapter 7 as the "emergency chapter" of the *Code,* fire pumps are covered in Article 695. It establishes requirements for electric power sources and switching and control equipment associated with fire pump drivers. However, it doesn't specify performance of fire pumps, which is covered by NFPA 20, *Standard for the Installation of Stationary Pumps for Fire Protection*.

Must Keep Operating

This is the central concept of Article 695. Because of their critical life-safety mission, fire pumps are intended to operate under conditions that would cause ordinary motor-driven equipment (wired to the requirements of Article 430) to be disconnected from its power supply and shut down. Fire pumps are intended to operate until a fire is extinguished, the pump is deliberately shut down, or the fire pump itself is destroyed. For this reason, many requirements of Article 695 are different from those elsewhere in the *NEC* for nonemergency equipment.

Reliable Power Source

Note

Appendix A of NFPA 20 provides the authority having jurisdiction with additional guidance for determining power supply reliability.

Fire pumps are required to have a reliable power source separate from other building equipment, as described in 695.3 and 695.4 (**Figure 4.9**). This power supply is required to be able to support the locked-rotor currents (i.e., maximum load under stalled conditions) of the fire pump motor and any associated equipment.

Overcurrent Protection

The fire pump overcurrent protective device must carry the locked-rotor currents indefinitely. In other words, the circuit breaker or fuse is not set to trip and protect the circuit conductors from overheating, as normally is the case under *NEC* rules. Instead, the circuit conductors are permitted to be damaged or destroyed if this will keep the fire pump operating longer to support firefighting efforts.

Conductors

Supply conductors must be sized to carry 125 percent of the fire pump motor full-load current rating. They are protected against short circuit only, and don't have overload protection. Fire pump supply circuit conductors must be kept completely separate from all other wiring and must be installed using wiring methods that provide a minimum 1-hour fire resistance rating. To provide protection from physical damage, all wiring from controllers to pump motors must be installed in metal raceways, Type MI cable, or Type MC cable with an impervious covering [695.6(E)].

Hazardous (Classified) Locations

The *National Electrical Code* uses two different systems to define hazardous locations. Articles 500–503 cover the traditional North American system using the terms *Class*, *Division*, and *Group*. Articles 505 and 506 describe the Zone system defined in IEC standards. (Article 504 describes the related protection concept of *intrinsically safe systems*.)

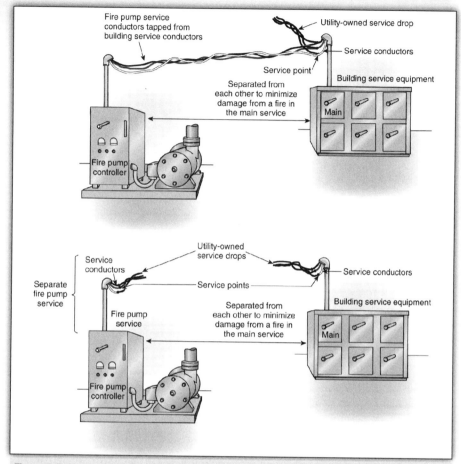

Figure 4.9 Two configurations are permitted by 695.3(A)(1) for supplying a fire pump from utility-owned service drops.

Classes, Divisions, and Groups

Classes describe types of atmospheres in which hot surfaces or electric sparks could potentially cause a fire or explosion:

- **Class I location (flammable gases, vapors, or liquids).** Typical Class I locations include refineries, oil drilling rigs, and chemical manufacturing plants.

- **Class II location (combustible dusts).** Typical Class II locations include grain elevators and coal-handling facilities.

- **Class III location (ignitible fibers and flyings).** Typical Class III locations include textile plants, paper mills, and sawmills.

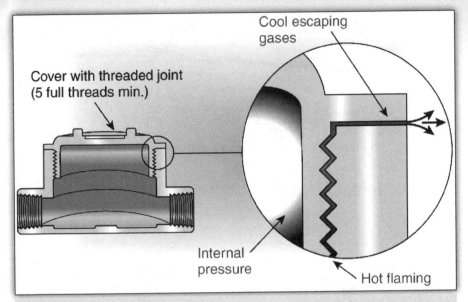

Figure 4.10 Hot gases are cooled as they pass through the threads of a screw-type cover of an explosion-proof junction box. Reprinted with permission from 2005 NEC Handbook, copyright © 2005, National Fire Protection Association.

Each of the three classes describing hazardous substances has two divisions, which describe degrees of hazard. Divisions describe whether an ignition hazard exists under *normal* operating conditions or only under *abnormal* ones:

- **Division 1 location (normal).** Division 1 locations have flammable substances present under normal operating conditions, and require the use of explosion-proof equipment to prevent the release of flames or hot gases at any time. **Figure 4.10** shows how hot gases from an internal explosion cool as they pass through the screw-threads of the junction box cover.

- **Division 2 location (abnormal).** Division 2 locations have flammable substances present only under *abnormal* conditions, such as the failure or rupture of equipment or a human operational error.

Class I is further subdivided into four groups—**Groups A, B, C,** and **D**—based on the specific gases, vapors, or dusts present. Equipment used in these atmospheres must be approved and marked for Class I and also the specific group involved.

Class II is divided into **Groups E, F,** and **G** based on the specific combustible dusts present. Equipment for use in these atmospheres must be approved and marked for Class II and also the specific group.

Intrinsically Safe Systems

Intrinsically safe apparatus, wiring, and systems operate at such low-energy levels that they can be installed in hazardous (classified) locations using general-purpose enclosures. However, conduits and cables still are required to be sealed under certain conditions to minimize passage of gases, dusts, and vapors.

Zone System

Zone classifications are an alternative to the class-division system for designating hazardous (classified) locations:

- **Class I, Zone 0, 1, and 2 locations.** These are more or less equivalent to Class I, Division 1 and 2 locations.
- **Zone 20, 21, and 22 locations.** These are more or less equivalent to Class II and Class III locations.

Areas with Flammable Fuels and Chemicals

NEC Articles 511–516 apply to specific locations where flammable fuels and chemicals are used.

- Article 511—Commercial Garages, Repair and Storage (doesn't apply to parking structures)
- Article 513—Aircraft Hangars
- Article 514—Motor Fuel Dispensing Facilities
- Article 515—Bulk Storage Plants
- Article 516—Spray Application, Dipping, and Coating Processes

General principles for electrical installations in these environments generally includes the following. See the individual articles listed previously for more details.

Class I. Due to the presence of flammable vapors that are either heavier or lighter than air, areas near the floor and near the ceiling are classified as either Class I, Division 1 or Class I, Division 2 (**Figure 4.11**). These classifications can be modified using ventilation systems, or by closing off specific areas, such as stockrooms and offices.

Wiring methods. Raceways passing through boundaries of Class I locations must be sealed. Fixed wiring above Class I locations must be installed in raceways or damage-resistant cable types such as MI, TC, and MC.

Arcing equipment and lighting. Arcing equipment and luminaires/lampholders located less than 12 feet above any Class I location must be totally enclosed or otherwise constructed to prevent the escape of sparks or hot-metal particles.

Emergency power cutoff. Circuit for fuel-dispensing and pumping equipment must be provided with a means to simultaneously disconnect all conductors, including the grounded conductor (**Figure 4.12**).

Health Care Facilities

NEC Article 517 covers electrical systems in health care facilities including hospitals, clinics, doctor and dentist offices, limited care facilities, nursing homes, and diagnostic

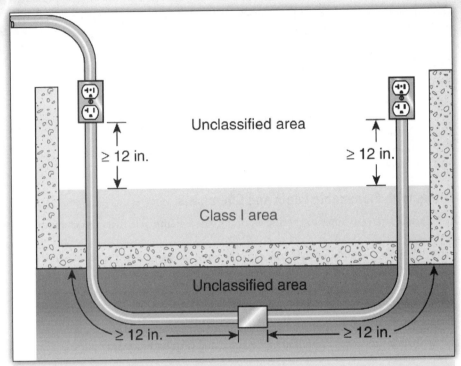

Figure 4.11 Seals aren't required for conduits that pass unbroken though a Class 1 hazardous location. Reprinted with permission from 2005 NEC Handbook, copyright © 2005, National Fire Protection Association.

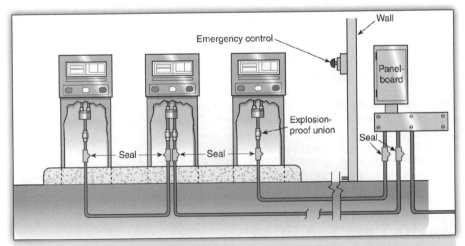

Figure 4.12 Installations of gasoline dispensers (pumps) are required to have sealing fittings where raceways enter a Class I location, plus an emergency disconnect switch.

imaging facilities. (It doesn't apply to veterinary facilities.) Article 517 has more "extracted" rules taken from a different NFPA document than any other article in the *Code*. The source document is NFPA 99, *Standard for Health Care Facilities*. Major safety concerns of Article 517 include the following:

Reliable Power Supply

Essential electrical systems. Part III of Article 517 defines *essential electrical systems* fed from normal and alternate sources, each of which is capable of supplying (by itself) a limited amount of lighting and power considered essential for life safety and operations under emergency conditions. Essential electrical systems are required for hospitals, nursing homes, and limited care facilities (sometimes called *urgent care facilities*), but not for doctor and dentist offices.

Selective ground-fault protection. When ground-fault protection for equipment (GFPE) is provided for services as required by 215.10 or 230.95, 517.17 defines special requirements for health care facilities. These requirements are intended to limit the extent of an outage due to GFPE operation.

Flammable Atmospheres

The use of oxygen in health care facilities increases the risk of fire from an electrical spark. Although health care facilities are not considered hazardous (classified) locations, Article 517 pays a great deal of attention to grounding and bonding.

Redundant grounding. Patient care areas must use metallic wiring methods (metal raceways or cables with metal sheaths) and have separate, insulated, copper grounding equipment conductors. These requirements create two parallel equipment grounding paths to minimize the risk of sparking or arcing in patient care areas that may have flammable or oxygen-rich atmospheres.

Shock Protection

The presence of moisture, and the use of electrical monitoring and treatment equipment, increases the risk of electrical shock in patient care areas of health care facilities. For this reason, GFCI protection for personnel is required in certain locations. Section 517.16 also permits use of *isolated ground receptacles* with insulated grounding terminals.

Medical X-Ray Installations

Article 517 covers electrical aspects of X-ray installations for medical diagnostic purposes. Industrial X-ray installations for purposes such as inspecting welds are covered by Article 660.

Manufactured Buildings and Mobile Homes

NEC Articles 545 and 550 cover electrical systems installed at the factory in prefabricated buildings, which then are moved to a site where they are erected or assembled. The electrical systems in such manufactured buildings can't be inspected by the AHJ during onsite construction, in the usual way.

Such prefabricated buildings may be used as dwellings, or for other purposes including construction trailers, clinics, bookmobiles, and mobile banks. Article 550 also includes the requirements for electrical distribution systems installed at mobile home parks.

Mobile Homes and Manufactured Homes

A mobile home is designed for use without a permanent foundation. A manufactured home usually is larger, may be transported in sections for assembly on site, and often has a permanent foundation. However, in the *NEC*, the general term *mobile home* includes both manufactured homes [550.2] and manufactured buildings used for nonresidential purposes [550.4].

Internal Electrical Systems

Most mobile homes have 120/240-volt, single-phase, 3-wire distribution systems similar to other types of dwellings. The rules for load calculations are in Article 555, and are similar to those for dwellings in Article 220. The rules for branch circuits, receptacle outlets, and GFCI/AFCI protection of receptacles also are similar to those for other dwellings in Article 210.

Marinas, Boatyards, and Floating Buildings

Proximity to water creates particular electrical safety concerns. Although boats and watercraft are outside the scope of the *NEC* [90.2(B)], Article 555 covers marinas and boatyards, while Article 553 covers floating buildings such as restaurants, bars, river casinos, and retail stores.

Floating buildings are moored in a permanent location and have a premises wiring system connected by permanent wiring to a source of electric supply. By contrast, boats have self-contained electrical systems and, when temporarily moored, are connected to shore power by means of attachment plugs and receptacles as described in Article 555 (which also covers floating piers). Key concepts include the following:

Flexible Wiring Methods

Both floating buildings and marina shore power feeders use liquid-tight flexible conduits or extra-hard usage cable listed for wet locations and sunlight resistance, so that motion of the water surface and changes in the water level won't cause unsafe conditions.

Electrical Datum Plane

At marinas, the electrical datum plane is a reference dimension above the normal highest water level [555.2]. Electrical equipment generally isn't installed, or connections made, below the electrical datum plane.

Shore Power

Special assemblies called *marine power outlets* are used to supply shore power to boats [555.2]. They include receptacles, overcurrent devices, watt-hour meters, and means for strain relief on power cords.

GFCI Protection

General-use receptacles installed in outdoor areas of marinas and boatyards, other than those used for shore power, are required to have GFCI protection for personnel. Those installed indoors must follow the rules of 210.8(A).

Gasoline-Dispensing Stations

Marine filling stations must also comply with Article 514.

Single-Family Dwellings Not Covered

Article 555 doesn't apply to private, noncommercial docking facilities at single-family dwellings. However, 218.10(A)(8) specifically requires ground-fault circuit-interrupter protection for receptacle outlets installed within boathouses, and 210.8(C) requires GFCI protection for outlets that supply boat hoists supplied by 125-volt, 15- and 20-ampere branch circuits in dwelling unit locations.

Recreational Vehicles, RV Parks, and Park Trailers

The *National Electrical Code* doesn't cover electrical installation in automotive vehicles other than mobile homes and recreational vehicles [90.2(B)]. Articles 550, 551, and 552 cover both the wiring and equipment in or on RVs (travel trailers, camping trailers, truck campers, and motor homes), and the electrical distribution systems of RV parks. Unlike mobile homes, which are intended for long-term or permanent habitation, RVs are intended only as temporary living quarters.

Types of Electrical Systems

Some RVs have 120- or 120/240-volt electrical systems and connect to an external power supply by means of receptacles rated 20, 30, or 50 amperes. Others have combination electrical systems that are powered from a 120-volt AC source and convert this to a lower DC voltage for utilization equipment.

Rules for load calculations, outlets, and branch circuits are generally similar to those for other dwellings (**Figure 4.13**). However, there is no requirement for AFCI protection of branch circuits serving receptacle outlets in recreational vehicles and park trailers.

Some RVs and trailers have low-voltage systems powered either from their own batteries or by plugging into an automobile's 12-volt socket. These aren't within the scope of the *NEC*; a FPN to 551.4(B) refers *Code* users to two other industry standards for information.

Temporary Installations

NEC Article 590 applies to temporary power and lighting installations for construction, demolition, repair, and similar activities (**Figure 4.14**). It doesn't cover the types of

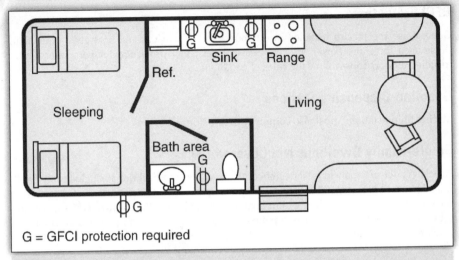

Figure 4.13 Recreational vehicle GFCI protection requirements are similar to those for permanent dwelling units [551.41(C)].

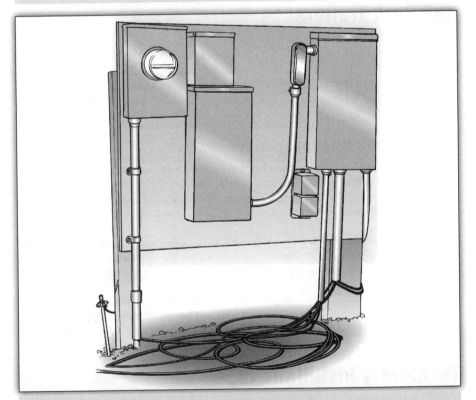

Figure 4.14 Temporary services used at construction sites provide electricity metering and overcurrent protection.

"traveling" temporary electrical systems described in Article 525. However, it does cover temporary decorative lighting installations for holidays, at bars and restaurants, and so on. Important considerations include the following:

Adequacy

Engineers who design buildings and similar structures should have a good understanding of temporary power systems. Specifying a system with adequate capacity and a sufficient number of outlets, and that provides good lighting and is well-maintained over the life of a construction project, often has a major impact on the productivity of construction trades.

For more detailed information, see NECA 200–2002, *Recommended Practice for Installing and Maintaining Temporary Electrical Power at Construction Sites* (ANSI).

Cable Wiring Methods

Temporary electrical systems generally use cable wiring methods, particularly nonmetallic-sheathed cable. Unlike the case with permanent buildings [334.10], there is no height or building-type limitation Type NM or NMC cable used in temporary installations [590.4(C)].

Ground-Fault Circuit-Interrupter Protection

All 125-volt, single-phase 15-, 20-, and 30-ampere receptacle outlets used to provide power for temporary installations are required to have GFCI protection for personnel [590.6]. This specifically includes receptacle outlets that are part of the permanent wiring of the building [590.6]. Listed portable GFCI devices are typically used for this purpose (**Figure 4.15**).

Decorative Lighting

Temporary electrical installations are permitted only for limited periods of time. In particular, 590.3(B) restricts temporary installations of holiday decorative lighting to a maximum of 90 days. The intent is to limit the use of cord-and-plug-connected "Christmas lights" widely used for decorative lighting at bars, restaurants, hotels, and shopping centers; along city streets; and in other outdoor locations.

Note

The 90-day limitation on plug-in strings of decorative lighting used outdoors is widely violated and poorly enforced.

This limitation on the use of such decorative lighting is in line with the *NEC*'s general intent to prevent the use of temporary wiring methods for more permanent applications. More permanent methods of decorative outdoor lighting are covered in:

- Article 410 — Luminaires, Lampholders, and Lamps
- Article 411 — Lighting Systems Operating at 30 Volts or Less

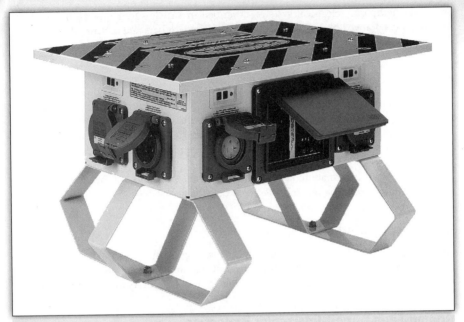

Figure 4.15 This type of power outlet unit with GFCI-protected receptacles is commonly used to provide temporary power on construction sites.

590.3 Time Constraint.

(A) During the Period of Construction. Temporary electrical power and lighting installations shall be permitted during the period of construction, remodeling, maintenance, repair, or demolition of buildings, structures, equipment, or similar activities.

(B) 90 Days. Temporary electrical power and lighting installations shall be permitted for a period not to exceed 90 days for holiday decorative lighting and similar purposes.

(C) Emergencies and Tests. Temporary electrical power and lighting installations shall be permitted during emergencies and for tests, experiments, and developmental work.

(D) Removal. Temporary wiring shall be removed immediately upon completion of construction or purpose for which the wiring was installed.

Chapter 6—Special Equipment

NEC Chapter 6 deals with a broad range of specialized equipment and contains more articles than *NEC* Chapter 4. Although Chapter 7 typically is considered the emergency systems chapter of the *Code*, fire pumps are included at the end of this chapter:

- Article 600—Electric Signs and Outline Lighting
- Article 604—Manufactured Wiring Systems

- Article 605—Office Furnishings (Consisting of Lighting Accessories and Wired Partitions)
- Article 610—Cranes and Hoists
- Article 620—Elevators, Dumbwaiters, Escalators, Moving Walks, Wheelchair Lifts, and Stairway Chair Lifts
- Article 625—Electric Vehicle Charging System
- Article 626—Electrified Truck Parking Space
- Article 630—Electric Welders
- Article 640—Audio Signal Processing, Amplification, and Reproduction Equipment
- Article 645—Information Technology Equipment
- Article 647—Sensitive Electronic Equipment
- Article 650—Pipe Organs
- Article 660—X-Ray Equipment
- Article 665—Induction and Dielectric Heating Equipment
- Article 668—Electrolytic Cells
- Article 669—Electroplating
- Article 670—Industrial Machinery
- Article 675—Electrically Driven or Controlled Irrigation Machines
- Article 680—Swimming Pools, Fountains, and Similar Installations
- Article 682—Natural and Artificially Made Bodies of Water
- Article 685—Integrated Electrical Systems
- Article 690—Solar Photovoltaic Systems
- Article 692—Fuel Cell Systems
- Article 695—Fire Pumps

Key Concepts

The 24 articles of *NEC* Chapter 6 contain both detailed requirements for installing electrical equipment and safety-related design and performance requirements that provide guidance for manufacturers and a basis for product listing. Chapter 6 is too wide ranging to summarize in a short space. Important concepts that engineers designing and specifying electrical installations must understand include the following. (Because audio installations have many similarities to low-voltage and limited-energy systems covered in *NEC* Chapters 7 and 8, Article 640 is included in Unit 5: Other Systems of this book.)

Elevators, Escalators, and Wheelchair Lifts

Many types of related equipment are covered by Article 620, which also shares common characteristics with Article 610 on cranes and hoists. Elevators, wheelchair lifts, and similar devices are characterized by intermittent use and short-time operation. For this reason, Table 620.14 provides demand factors for calculating elevator feeder loads.

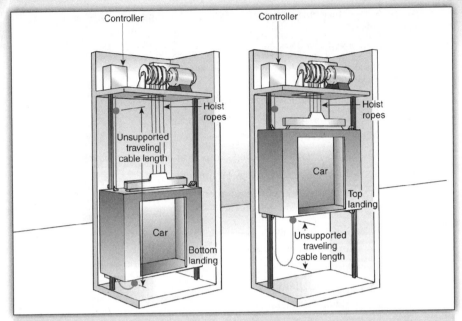

Figure 4.16 Traveling cables are used for power, communications, and control circuits of elevators.

Wiring Methods

Equipment covered by Article 620 uses a number of special wiring techniques. These include traveling cables in hoistways (**Figure 4.16**), extensive use of cables and flexible raceways to minimize the transmission of noise and vibration at connections to motorized equipment, and use of conductors smaller than 24 AWG for low-voltage electronic controls with correspondingly low currents.

Other Equipment and Wiring Prohibited

Elevator hoistways make convenient pathways for running electrical cables and raceways from the basement to the roof of a building. To prevent the installation of equipment that might interfere with the safe operation of elevators, and to prevent unqualified persons from working in these areas, 620.37 permits only wiring and equipment directly related to elevator functions to be installed within hoistways and machine rooms. Similar prohibitions on installing other types of wiring and equipment in elevator hoistways and machine rooms are found throughout the *Code*.

Information Technology Equipment

NEC Article 645 covers equipment, wiring, and grounding systems in computer rooms, as defined in NFPA 75, *Standard for the Protection of Information Technology Equipment*. It doesn't cover desktop computers in homes and offices, point-of-sale terminals in stores, security systems controllers, and other kinds of information technology equipment not located in dedicated computer rooms.

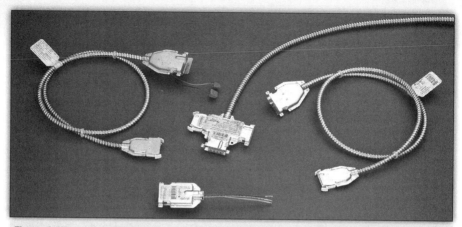

Figure 4.17 Polarized receptacles and connectors of a manufactured wiring system are required to be part of a listed assembly for the system.

Disconnecting Means

A means must be provided to disconnect all power to equipment within the information technology room, shut down dedicated HVAC systems serving the room, and close all fire/smoke dampers [645.10].

Flexible Cords and Cables

Flexible cords and cables are widely used for supplying power to information technology equipment and for interconnecting data-processing units. Where they run on the surface of the floor, flexible cords and cables must be protected against physical damage.

Wiring Under Raised Floors

Branch-circuit conductors under raised floors must be in metal raceways or cable types with metallic sheaths (AC, MC, or MI). Data cables installed under raised floors must be listed Type DP, which has fire-resistant characteristics. However, cords and attachment plugs can be used to connect equipment to branch circuits installed under raised floors of information technology equipment rooms.

Manufactured Wiring Systems and Office Furnishings

Manufactured Wiring Systems

Article 604 covers field-installed wiring systems that use premanufactured components. This saves installation time in the field and also permits repairs and changes after the initial installation to be made more easily. The most common manufactured wiring systems use factory-assembled cables with special plugs at each end to connect together fluorescent luminaires installed in suspended ceilings (**Figure 4.17**). The luminaires are factory equipped with matching polarized, locking-type receptacles.

Office Furnishings (Wired Partitions and Lighting Accessories)

Article 605 covers wired partitions and lighting components used to assemble office cubicles in open-plan areas (often these are called *office furniture*). They typically provide users with 125-volt, single-phase, 15- and 20-ampere receptacles; telephone and data outlets; and luminaires. In some cases, branch-circuit wiring and outlets are installed in partitions at the factory. In others, wiring channels are provided along the tops or bottoms of the partitions for field installation of conductors.

Telepower Poles

These vertical wiring channels also used in open-plan office environments to provide power and data outlets are not within the scope of either Article 604 or 605. Instead, they are considered a wiring method and must comply with the installation rules of Article 380—Multioutlet Assembly, along with similar horizontal channels commonly known by the proprietary names Wiremold® and Plugmold®.

Swimming Pools, Fountains, and Similar Installations

Article 680 covers construction of, installation of electrical wiring to, and metallic auxiliary equipment such as pumps and filters for all types of pools, fountains, spas, and hydromassage tubs. Its requirements apply whether this equipment is permanently installed or storable, and whether located indoors or out.

Article 680 applies to pools and similar installations for recreational, decorative, or therapeutic uses. Other bodies of water with associated electrical equipment, such as fish farm ponds, water treatment ponds, and aeration fountain pumps in lakes, are covered by Article 682. It has similar, but more limited, safety requirements.

Shock Protection

Because electric shock danger is increased by the presence of water, Article 680 pays a great deal of attention to safe installation of receptacles and luminaires; clearances between water and pool equipment, luminaires, clearance for overhead conductors (**Figure 4.18**); GFCI protection for personnel, and bonding.

How *NEC* Performance Requirements Influence Equipment Design

Many of the *NEC* articles discussed in this unit contain construction requirements for electrical equipment and systems. However, strictly speaking, these are performance requirements; the actual, detailed product design requirements are included in the product standards listed in this book's appendix. Although the *Code* isn't a design standard, it has a strong impact on the design of electrical products and systems through a three-step process:

1. *Safety problem identified.* A safety problem with current *NEC* rules is brought to the attention of the responsible Code-Making Panel (CMP), typically through

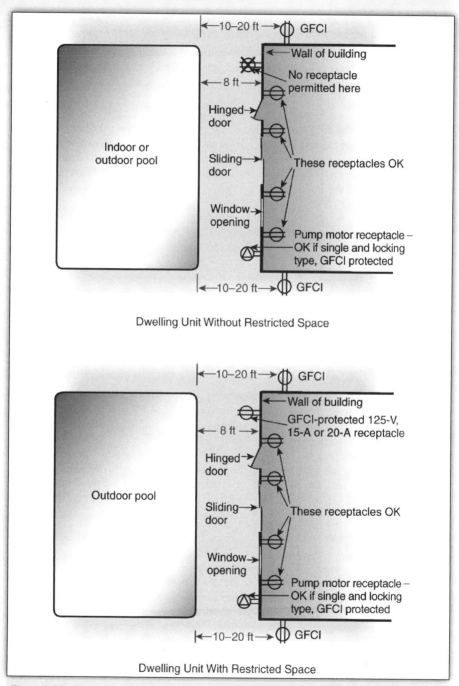

Dwelling Unit Without Restricted Space

Dwelling Unit With Restricted Space

Figure 4.18 All 15- and 20-ampere, 125-volt outdoor receptacles located within 20 feet of the inside walls of a pool or fountain are required to have GFCI protection, per 680.22(A)(5).

submission of public change proposals every 3 years. It also might happen through a request for a tentative interim amendment (TIA) or formal interpretation.

2. *New NEC performance requirement.* Depending on the nature of the safety problem, the CMP may specify a new installation technique, a new construction requirement for an existing type of equipment, a wholly new product, or a combination of all of these.

3. *New product standards created.* When a new edition of the *NEC* is published with the new requirement(s), product testing and certification agencies such as Underwriters Laboratories, Inc., develop standards to implement and verify them, and product manufacturers begin producing new products that meet the new requirements. (Product standards referenced in the *Code* are listed in the appendix.)

Typically, this takes time. Also, there may be existing products in the supply "pipeline" that meet the previous edition of the *Code* but not the new one. In such cases, 90.4 gives the authority having jurisdiction the ability to continue approving alternate products or methods for a limited time.

No Change Needed

Alternately, the CMP may decide that the current *National Electrical Code* rule is adequate to ensure safety and that new requirements aren't needed. Similarly, the CMP may decide that the proposed revision does not involve a safety issue. Many change proposals primarily involve design, esthetic, or convenience issues.

However, it is important to remember that the *Code* is a safety document that specifies safety requirements and not performance or quality attributes. The most common reason used by CMPs to reject proposed revisions of the *NEC* is lack of a convincing rationale that the proposed change will improve safety.

Lastly, a CMP may decide to limit or prohibit the use of an existing installation practice or type of equipment if the panel members can't find a way to create *Code* rules ensuring that electrical installations will be "essentially free from hazard" as required by 90.1(B).

Conclusion

This unit provides an overview of *NEC* safety rules primarily related to all types of electrical equipment. Although the *Code* classifies these in a variety of ways—as equipment for general use, special equipment, equipment for use in special occupancies, equipment forming a part of wiring methods, and so forth—all electrical equipment shares a number of common concerns under the *National Electrical Code*.

UNIT 5

Other Systems

Introduction

The *National Electrical Code* contains installation rules for all kinds of electrical products and systems. Many electrical professionals, including engineers and designers, think of the *Code* primarily as a "power wiring" book—and so it is—but not just power wiring. The *NEC* also contains detailed safety requirements for installation of all types of communications, signaling, and control systems. This unit covers the following:

- Low-Voltage and Limited-Energy *Code* Articles
- Chapter 7—Special Conditions
- Chapter 8—Communications Systems
- Other Low-Voltage Articles
- Low-Voltage Wiring: An Overview
- No *NEC* Rules for Carrier-Current Systems

A primary reason that many users don't realize the *Code* also applies to low-voltage and limited-energy systems is that many jurisdictions don't require permits for this work, and no permits means no inspection, as explained in Unit 1. However, there are important safety reasons to design low-voltage installations in compliance with *Code* rules:

- *Audio.* Audio voltages can be as high as 70 volts AC.
- *Telephone.* Telephone ringing voltages can be as high as 90 volts AC.
- *Shock hazard.* Incorrectly installed low-voltage wiring may accidentally become energized at line voltages, endangering both installers and users.
- *Grounding and bonding.* Proper grounding and bonding of communication circuits, CATV cables, TV and satellite masts, and the like are essential to prevent fires and electric shock from dangerous potential differences between the electrical systems.

Low-Voltage and Limited-Energy *Code* Articles

Wiring for audio, telephone, CATV, security, satellite dishes, fiberoptics, low-voltage lighting, and similar systems are within the scope of the *NEC*. Chapters 7 and 8, along with several other articles, specifically cover all these types of equipment and wiring.

These "lost" low-voltage articles of the *National Electrical Code* are the following; a brief summary of each article follows (**Table 5.1**).

Chapter 7—Special Conditions

Chapter 7 of the *National Electrical Code* covers both equipment and wiring systems that can be classified as representing special conditions, including emergency and standby power installations. The articles most concerned with low-voltage and limited-energy systems include the following:

Article 700—Emergency Systems

Emergency system wiring must be run separately from all other wiring [700.9(B)] and have special protection against fire to ensure that building emergency systems continue operating for as long as possible under fire conditions (**Figure 5.1**). There are no special wiring

Table 5.1 Low-Voltage *Code* Articles

Chapter 7—Special Conditions	Article 720—Circuits and Equipment Operating at Less than 50 Volts Article 725—Class 1, Class 2, and Class 3 Remote-Control, Signaling and Power-Limited Circuits Article 770—Optical Fiber Cables and Raceways
Chapter 8—Communications Systems	Article 800—Communications (Telephone) Circuits Article 810—Radio and Television Equipment Article 820—Community Antenna Television and Radio Distribution Systems Article 830—Network-Powered Broadband Communications Systems
Other Low-Voltage Articles	Article 411—Lighting Systems Operating at 30 Volts or Less Article 504—Intrinsically Safe Systems Article 640—Sound-Recording and Similar Equipment

requirements for legally required standby systems [Article 701] and optional standby systems [Article 702].

Article 720—Circuits and Equipment Operating (Less than 50 Volts)

This short article covers AC and DC power installations operating at less than 50 volts. Its original purpose was to establish safety rules for low-voltage farm lighting powered by batteries, so this article is rarely used today.

Article 725—Remote-Control, Signaling, and Power-Limited Circuits

This article covers remote-control, signaling, and power-limited circuits that are not an integral part of a device or appliance. These include systems such as low-voltage control, signaling, burglar alarms, and power-limited computer network cabling not within an information technology (computer) room covered by Article 645.

- *Signaling circuits* supply energy to an appliance or device that gives a visual and/or audible signal. Examples are doorbells, fire or smoke detectors, and burglar alarms.

- *Remote-control circuits* control other circuits or equipment. Examples are low-voltage relay switching of lighting (**Figure 5.2**), or remote control of motor loads.

- *Power-limited circuits* are used for functions other than signaling or remote control, but in which the source of the energy supply is limited to specified maximum power levels (expressed in volt-amperes). Low-voltage lighting in which 12-volt fixtures are fed from 120/12-volt transformers is a typical power-limited circuit application.

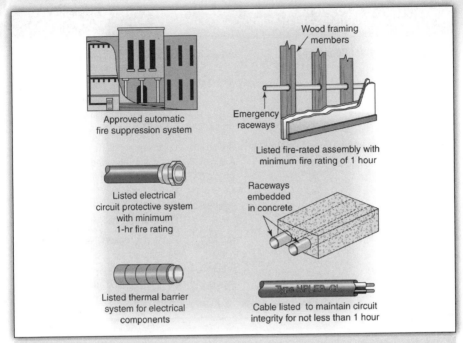

Approved automatic
fire suppression system

Wood framing
members

Emergency
raceways

Listed fire-rated assembly with
minimum fire rating of 1 hour

Listed electrical
circuit protective system
with minimum
1-hr fire rating

Raceways
embedded
in concrete

Listed thermal barrier
system for electrical
components

Type NPLFP-CI

Cable listed to maintain circuit
integrity for not less than 1 hour

Figure 5.1 Section 700.9(D)(1) specifies six techniques for protecting emergency system feeder and circuit wiring.

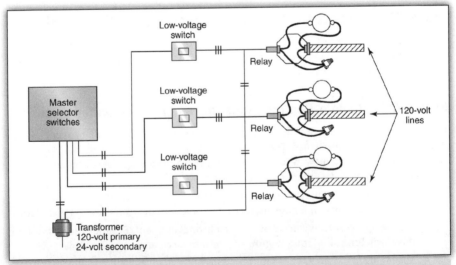

Low-voltage
switch

Relay

Low-voltage
switch

Relay

Master
selector
switches

120-volt
lines

Low-voltage
switch

Relay

Transformer
120-volt primary
24-volt secondary

Figure 5.2 Low-voltage lighting control is a typical application for Class 2 circuits.

Article 725 classifies these types of circuits into Class 1, 2, and 3 wiring systems. Class 1 includes all signaling and remote-control circuits that do not have the low current limitations of Class 1 and 3 circuits. Class 1 circuits operate at up to 600 volts, and are similar in many ways to line-voltage electric lighting and power circuits. Class 1 circuits must be installed in accordance with the appropriate articles of Chapter 3 [725.26]. Class 2 and Class 3 circuits are those in which the current is limited to certain specified low values by fuses or circuit breakers, by power supplies which deliver only small currents, or by other approved means. In general, Class 2 and 3 circuits aren't permitted to share the same raceway or enclosure with Class 1 or electric lighting and power circuits [725.55(A)].

Article 760—Fire Alarm Signaling Systems

This article covers the installation of wiring and equipment for fire alarm signaling systems. Nonpower-limited fire alarm (NPLFA) circuits can operate at up to 600 volts. Power-limited fire alarm (PLFA) circuits operating at lower voltages are supplied from a listed PLFA transformer, listed Class 3 transformer or power supply, or listed equipment marked to identify the PLFA power source; many PLFA systems operate at 24 VDC. While *NEC* Article 760 covers wiring methods for fire alarm systems, design and performance requirements for fire alarm systems are specified in NFPA 72, *Fire Alarm Code*. This is discussed further in Unit 6 of this book.

Article 770—Optical Fiber Cables and Raceways

This article covers the use of fiberoptic cables and listed optical fiber raceways used in stand-alone applications or as innerduct [770.12(B)]. It also covers composite cables (often called "hybrid" in the field) that combine optical fibers with current-carrying metallic conductors. These composite cables are classified as electrical cables (such as Type MC) under other articles of the *NEC* [770.9].

Chapter 8—Communications Systems

Chapter 8 of the *National Electrical Code* covers communications systems. It is independent of the other *Code* chapters except where they are specifically referenced in the following articles [90.3]:

Article 800—Communications Circuits

This article covers telecommunications systems. By extension, it also covers unshielded twisted-pair (UTP) cabling installed for computer networks (**Figure 5.3**), which the *NEC* and other industry standards consider a subset of telecommunications; see the FPN to 800.24. Communications wires and cables must have a minimum voltage rating of 300 volts to coordinate with the ratings of listed primary protectors [800.90(A), 800.179].

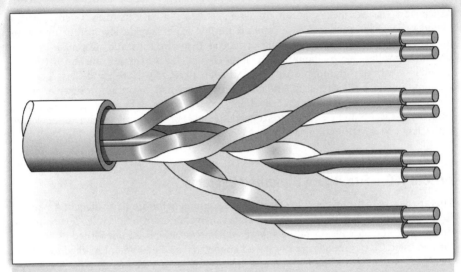

Figure 5.3 Computer networks wired using category-rated unshielded twisted-pair (UTP) cables are covered by Article 800.

Article 810—Radio and Television Equipment

Despite its title, this article primarily covers installation of antennas and lead-in conductors for radio and television receiving equipment. This includes TV antennas and satellite dishes, amateur radio (HAM) transmitting and receiving equipment, and installations at commercial radio and television broadcasting stations.

Article 820—Community Antenna Television and Radio Distribution Systems

This article covers coaxial cable installations for distributing radio-frequency signals, such as CATV installations in high-rise buildings. By extension, it also covers coaxial cable installed for other applications, such as computer networks and closed-circuit television (CCTV) cameras.

Article 830—Network-Powered Broadband Communications Systems

This article covers interactive broadband communications systems that deliver power to the receiving equipment over the center conductor of coaxial cable (**Figure 5.4**). Major concerns of Article 830 include power levels, cable types, and bonding and grounding.

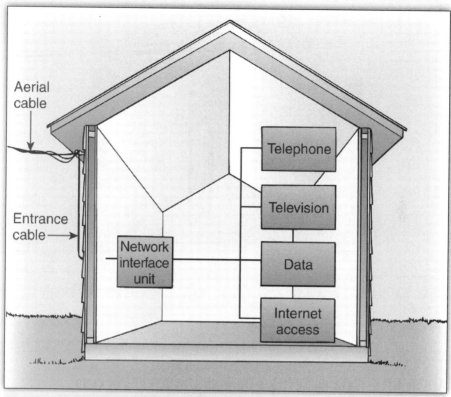

Figure 5.4 Network-powered broadband communications systems covered by Article 830 are sometimes called *multimedia systems.*

Power levels. Compared with other communications systems covered by the *Code*, network-powered communications systems operate at relatively high power levels. Table 830.15 indicates that low-power systems can be rated up to 100 volt-amperes at 100 volts and that medium-power systems can be rated up to 100 volt-amperes at 150 volts. The argument can be made that Article 830 doesn't really deal with low-voltage systems in the same sense as Articles 725 or 800.

Other Low-Voltage Articles

Other requirements for low-voltage and limited energy systems are scattered throughout the *Code*, but many are very specialized, such as rules for battery-powered DC systems in recreational vehicles and park trailers [Articles 551 and 552] or low-voltage wiring of medical X-ray equipment [Article 517]. The low-voltage articles of primary interest to engineers designing and specifying premises wiring systems are the following:

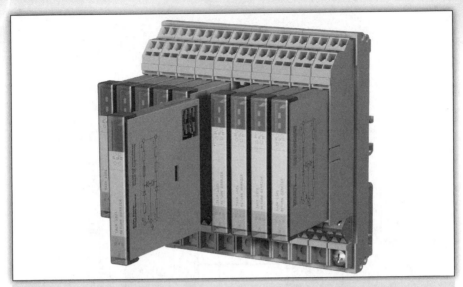

Figure 5.5 Intrinsic safety barriers limit the electrical energy available in a hazardous (classified) location.

Article 411—Lighting Systems Operating at 30 Volts or Less

This article covers listed lighting systems consisting of an isolated power supply operating at 30 volts RMS or less, with one or more secondary circuits limited to 25 amperes supplying luminaires. Such systems typically use 12-volt luminaires fed from 120/12-volt transformers. Lighting systems covered by Article 411 are used both outdoors (for landscape lighting) and indoors (for decorative lighting in places such as restaurants, stores, bars, and hotels).

Article 504—Intrinsically Safe Systems

Intrinsically safe apparatus, wiring, and systems are used in Class I, II, and III hazardous (classified) locations. The energy levels of these circuits are so low that any spark or heat generated is incapable of igniting a flammable or combustible atmosphere (**Figure 5.5**). Intrinsically safe systems are also discussed in Unit 4 of this book.

Article 640—Audio-Signal Processing, Amplification, and Reproduction Equipment

This article covers amplifiers, paging-music systems used in many types of buildings, audio recording and playback systems, and electronic musical instruments. It includes audio systems such as nurse call, live-sound recording, and temporary installations for events such as theatrical performances [Article 520], circuses and fairs, and outdoor concerts [Article 525]. In general, Article 640 installations use Class 2 and Class 3 wiring defined in Article 725.

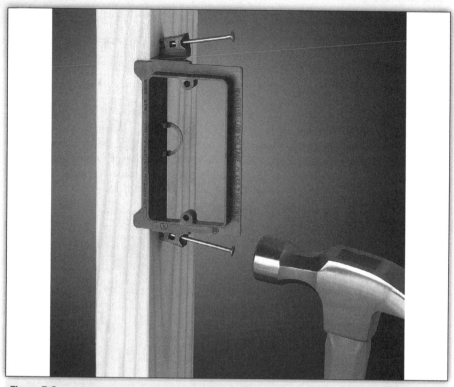

Figure 5.6 Communications (telephone) outlets are typically installed in nonmetallic brackets attached to studs. The *NEC* doesn't require boxes for low-voltage outlets.

Low-Voltage Wiring: An Overview

In addition to Chapter 3 wiring methods, *NEC* Chapters 7 and 8 contain rules for low-voltage and limited-energy systems. This section summarizes the differences between *Code* rules for power wiring systems (in Chapter 3) and for low-voltage and limited-energy systems (in Chapters 7 and 8).

Different Hazards

Because low-voltage and limited-energy systems operate at lower energy levels than power systems, the level of hazard is correspondingly lower. Less protection is needed to guard people against shock and electrocution hazard from low-voltage and limited-energy systems.

Raceways and Boxes Not Required

Generally speaking, low-voltage and limited-energy conductors aren't required to be installed in raceways; terminations and splices aren't required to be installed in boxes. Low-voltage devices such as telephone jacks and cable television outlets normally are installed on brackets, rather than in boxes (**Figure 5.6**).

Listing

Because they are typically installed without protective raceways and boxes, the *NEC* requires that low-voltage and limited-energy conductors be listed [725.7, 760.7, 770.21, 800.21, 725.82, 760.81, 770.179, 800.179(I), 820.179, and 830.179].

Supporting

Low-voltage and limited-energy conductors aren't required to comply with 300.11(A) requiring raceways, cables, boxes, and similar items to be securely fastened in place independently of the support wires for ceiling systems. However, a number of low-voltage and limited-energy articles have sections on mechanical execution of work that require cables and conductors to be installed in a neat and workmanlike manner and to be supported by the building structure.

- 640.6—Audio Systems
- 640.22—Audio Systems
- 640.23—Audio Systems
- 725.8—Class 1, 2, and 3 Circuits
- 760.8—Fire Alarm Systems
- 770.24—Fiber Optics
- 800.24—Communication Circuits
- 820.24—CATV (Coaxial Cable)
- 830.24—Broadband Systems

Abandoned Cables

Accessible portions of abandoned low-voltage cables are required to be removed, to prevent the accumulation of unused cables that increase fuel load in the event of fire, and to minimize congestion in spaces such as the areas above suspended ceilings. Abandoned cable typically is defined as installed cable that is not terminated at equipment and not identified for future use with a tag [725.3(A), 760.3(A), 770.3(A), 800.3(A), 820.3(A), 830.3(A)].

Suspended Ceiling Panels

Low-voltage and limited-energy conductors cannot be installed in such a way that they prevent the removal of panels that provide access to electrical equipment, including suspended ceiling panels [725.7, 760.7, 770.21, 800.21, 820.21, 830.21]. This means that low-voltage and limited-energy conductors must be properly supported above suspended ceilings (by cable trays, for example) and not simply laid across ceiling panels.

Plenum- and Riser-Rated Cables

Because they are typically installed without protective raceways and boxes, the *NEC* requires that low-voltage and limited-energy conductors installed in air-handling spaces and in vertical risers that run from floor to floor be listed for the application. The intent is to reduce the

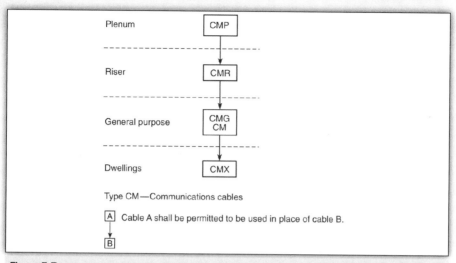

Figure 5.7 Cable substitution hierarchy. Reprinted with permission from 2005 NEC Handbook, copyright © 2005, National Fire Protection Association.

spread of products of combustion (flame, smoke, and gases) throughout a building by using less-flammable cables and conductors in these spaces.

Plenum cables have a letter "P" in their designation, and riser cables have a letter "R" in their designations [725.61, 760.30(B), 760.61, 760.81, 770.154, 800.154, 820.154, 830.154]. Cable types with greater fire resistance are permitted to be substituted for cable types with lower fire resistance (**Figure 5.7**).

Separations from Power Wiring

In general, the *Code* requires that low-voltage power wiring be run in separate raceways, boxes, and other enclosures than power wiring—or separated from it by a barrier [725.55, 770.133, 800.133, 820.133, 830.133]. Combination power and communications wall outlets have boxes with internal barriers between the low-voltage wiring and power conductors. Likewise, surface-mounted raceways and wiring channels are available with different color-coded compartments, one for line-voltage conductors, and one for low-voltage.

No *NEC* Rules for Carrier-Current Systems

None of the low-voltage and limited-energy *Code* articles just summarized deal specifically with powerline- and phoneline-carrier technologies—such as X-10 Powerhouse, CEBus, and HomePNA—that transmit control and communications signals over existing conductors originally installed for other purposes. This is because the *NEC*, as a safety code, doesn't contain design or performance requirements not related to protecting people and property from electrical hazards. *NEC* 810.1 specifies that equipment and antennas used for coupling carrier current to powerline conductors aren't within the scope of Article 810.

Conclusion

The *National Electrical Code* contains safety rules for all types of electrical installations, including communications, signaling, and control systems. Although many electrical professionals think of the *Code* primarily as a "power wiring" book, low-voltage and limited-energy systems also must be installed properly to comply with the *NEC's* stated purpose: "The practical safeguarding or persons and property from hazards arising from the use of electricity [90.1(A)]."

UNIT 6

How the *NEC* Is Related to Other Codes and Standards

Introduction

Although the *National Electrical Code* is written as a self-contained book of electrical safety rules, many other industry codes and standards also contain mandatory requirements that affect electrical installations. Most, but not all, of these are published by the National Fire Protection Association (NFPA). Some of these other industry codes and standards are adopted for regulatory use by states, cities, and counties. Some are referenced in customer requests for proposal (RFPs) or corporate specifications and design guidelines.

Engineers must be aware of the "interlocking" requirements of these other codes and standards requirements that supplement the wiring rules of the *National Electrical Code*, when designing and specifying electrical installations. This unit covers the following:

- U.S. Electrical Safety System
- *NEC* References to Other Codes and Standards
- NFPA Building Codes and Other Regulatory Standards
- Horizontal and Vertical Standards
- NFPA 70E-2004, *Standard for Electrical Safety in the Workplace*®
- International Code Council (ICC) Publications

U.S. Electrical Safety System

Electrical construction is a business defined by codes and standards. The United States is unique in its reliance on voluntary industry standards to regulate health and safety. There are many mandatory government regulations that affect electrical installations and product design, including those of the Department of Energy, the Environmental Protection Agency (EPA), and the Occupational Safety and Health Administration (OSHA).

Most U.S. electrical codes and standards, however, are developed by private-sector organizations such as engineering societies and industry trade associations. They become regulatory documents when states, counties, and cities adopt them by reference as laws. The resulting public-private regulatory framework is often called the U.S. electrical safety system. Private-sector organizations that develop voluntary standards are known by two interchangeable terms: standards developing organizations (SDOs) and accredited standards developers (ASDs).

The U.S. government recognizes the importance of this voluntary standards system for meeting federal procurement and safety goals through the Office of Management and Budget Circular A-119 and the 1996 Technology Transfer Act. Both of these encourage government agencies to depend on the private standards setting to the extent feasible and give guidance for agency participation in the voluntary standards system. Federal participation is coordinated by the National Institute of Standards Technology (NIST), working in close coordination with the American National Standards Institute (ANSI), the federation of U.S. standards developing organizations.

The U.S. electrical safety system consists of many interrelated codes and standards. Most of them are developed under ANSI-accredited procedures. These codes and standards can be logically grouped into three "tiers":

1. *First Tier:* Mandatory Installation Codes (Building Codes)
2. *Second Tier:* Product Standards
3. *Third Tier:* Optional Standards

First Tier: Mandatory Installation Codes (Building Codes)

Among U.S. building codes, the two primary ones governing electrical construction are the *National Electrical Code* (ANSI/NFPA 70-2008) and the *National Electrical Safety Code* (ANSI/IEEE C2-2007). Though their titles are confusingly similar, the two documents have different scopes and can actually be considered as mutually exclusive:

- The *National Electrical Code (NEC)* governs electrical installations in buildings and similar structures, generally operating at 600 volts or less.
- The *National Electrical Safety Code* (*NESC*) governs electric utility and heavy industrial installations, typically operating at thousands of volts.

The relationship between the *NEC* and the *NESC* was described in more detail in Unit 2 of this book. Examples of other ANSI-approved codes with requirements that affect electrical installations include the following:

- ANSI/NFPA 5000-2003, *Building Construction and Safety Code,* defines building occupancies and methods of fire-resistive construction, including firestopping of penetrations for electrical raceways and other components.
- ANSI/NFPA 101-2000, *Life Safety Code,* specifies locations where exit signs and emergency lighting are required.
- ANSI/NFPA 72-2007, *National Fire Alarm Code®,* provides requirements for fire alarm systems and components. It addresses installation as well as testing and maintenance.

Both NFPA 101 and NFPA 72 refer to the *NEC* for wiring rules related to these types of safety-related lighting and fire protection equipment.

Second Tier: Product Standards

The *NEC* and other building codes are primarily installation documents, but they also contain some basic performance and construction requirements for electrical equipment within their scopes. Conversely, detailed construction and operational specifications for electrical equipment, often referred to as *product standards*, sometimes contain installation requirements.

The *National Electrical Code* contains informational references to nearly 200 ANSI-approved product standards, more than 90 percent of them published by Underwriters Laboratories Inc. However, the *NEC* has no product approval mechanism of its own. Electrical construction professionals such as electrical contractors, electricians, consulting engineers, and electrical inspectors accept the UL mark as evidence that a product (such as a cable, lighting fixture, or circuit breaker) complies with the minimum safety-related performance requirements of the *NEC*. Product standards can be seen as complementary to the *NEC*.

Because high-voltage utility wiring products are not, for the most part, certified or listed by independent testing agencies, the *National Electrical Safety Code* contains references to fewer ANSI-approved product standards (just 29) and uses them in a different way. The *NEC* contains no mandatory references to other standards, including those published by NFPA. However, Section 3 of the *NESC* states that "The following standards form a part of the *National Electrical Safety Code* to the extent indicated in the rules herein."

Third Tier: Optional Standards

The remaining category contains third-tier electrical standards, which aren't adopted for regulatory use by state and local governments. Instead, they are commonly referenced in a consulting engineer's design specifications or included in a customer's RFP or electrical construction guidelines.

Third-tier standards supplement installation codes by providing additional requirements in areas such as construction quality, energy conservation, system testing and commissioning, and maintenance. None of these factors necessarily affect the minimum safety of electrical installations, yet all of them can strongly influence the long-term cost and usability of electrical systems to their owners: industrial concerns, store and restaurant chains, public agencies such as school and library boards, individual homeowners, investors, and so forth.

There are some common reasons that an electrical system's designers or customers might specify third-tier standards:

- Enhanced safety (e.g., requiring safety minimums beyond those of first-tier installation codes)
- User comfort and convenience (e.g., requiring more light fixtures or wall switches)
- Lower costs (e.g., requiring energy-efficient motors and lamps, or occupancy sensors that turn on lights only when a room is occupied)
- Improved performance (e.g., requiring oversized, heavy-duty, or premium-quality equipment, which typically lasts longer than electrical equipment designed to performance minimums)

There are so many of these ANSI-approved optional electrical standards that any list risks being incomplete. At a minimum, it would include standards from the following organizations:

- *Institute of Electrical and Electronics Engineers (IEEE) (www.ieee.org)*—electrical product standards, recommended system design guides such as the IEEE Color Book series of publications
- *National Electrical Contractors Association (NECA) (www.neca-neis.org)*—installation standards intended for reference in project plans and specifications. Because they deal with electrical construction performance and quality, these National Electrical Installation Standards (NEIS) in some cases contain requirements over and above the minimum safety rules of the *NEC* and *NESC*.
- *National Electrical Manufacturers Association (NEMA) (www.nema.org)*—over 200 construction standards, rating standards, and application guides for electrical construction products

- *Illuminating Engineering Society of North America (IESNA) (www.iesna.org)*—recommended lighting design practices, testing and rating standards for light sources
- *Telecommunications Industry Association (TIA) (www.tiaonline.org)*—design, performance, and testing standards for telecommunications installations, including computer networking applications

NEC References to Other Codes and Standards

The *NEC* is a stand-alone safety code that does not contain mandatory references to other codes and standards. However, in order to cover many different electrical installation situations, it depends on technical material contained in other codes and standards. It uses two methods to make these technical requirements available to *Code* users:

1. Mandatory extracts from other NFPA documents
2. Informational fine print note (FPN) references to documents published by NFPA and other organizations

Extracts from Other NFPA Documents

The *NEC* contains a number of rules taken verbatim from other NFPA codes and standards. In these cases, the "extracted" material goes through the consensus process to develop the *Code* and becomes an integral part of it. Such extracted material can be either mandatory or permissive *NEC* rules, depending on the type of language involved.

Extracted language from other NFPA documents in the *NEC* is identified to show its source. However, this doesn't mean that other rules in the source standard also become part of the *Code*. Only the extracted material that appears in the document is an enforceable part of it.

The following *NEC* material is extracted from NFPA 30A, *Motor Fuel Dispensing Facilities and Repair Garages*. Please note that although the *NEC Style Manual* prohibits using the term "can," it appears in this extract because other NFPA standards use this term.

514.3 Classification of Locations

(A) Unclassified Locations. Where the authority having jurisdiction can satisfactorily determine that flammable liquids having a flash point below 38ºC (100ºF), such as gasoline, will not be handled, such location shall not be required to be classified **(Figure 6.1)**.

FPN References to Other Codes and Standards

The *NEC* contains many FPNs that refer to other documents. These are purely informational, because FPNs are not enforceable *Code* requirements. They also are very general, because FPNs refer to entire publications only, and not to individual chapters or sections

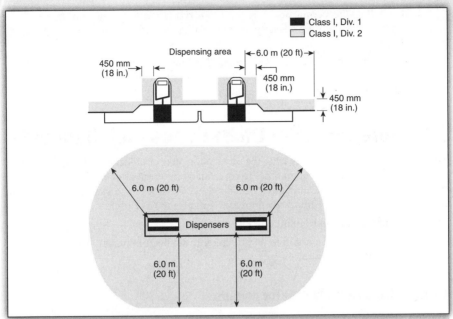

Figure 6.1 Classified areas adjacent to dispensers as detailed in Table 514.3(B)(1) [NFPA 30A: Figure 8.3.1]. Reprinted with permission from 2005 NEC Handbook, copyright © 2005, National Fire Protection Association.

within those publications. Because FPN references aren't mandatory, engineers designing and specifying electrical installations are required to follow the referenced standards only if those standards have been adopted for regulatory use by a state or local government and are included as mandatory provisions of a client's RFP or construction guidelines.

NFPA Standards Adopted for Regulatory Use

A number of NFPA codes and standards are frequently adopted for regulatory use by states, cities, and counties:

- NFPA 1, *Uniform Fire Code*
- NFPA 72, *National Fire Alarm Code*
- NFPA 99, *Health Care Facilities*
- NFPA 101, *Life Safety Code*
- NFPA 780, *Lightning Protection*

Horizontal and Vertical Standards

NFPA codes and standards contain a number of electrical requirements that supplement *National Electrical Code* rules, and other nonelectrical rules that affect the application of the *NEC*.

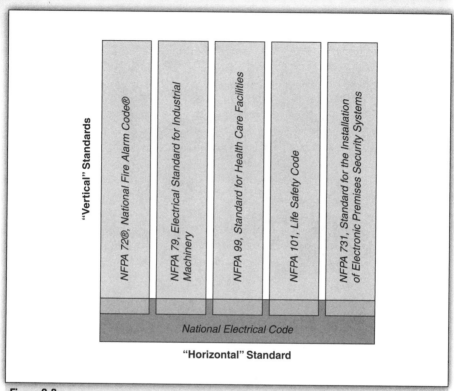

Figure 6.2 The *NEC* applies broadly to many types of installations. Other NFPA standards supplement the *NEC* with additional rules for specialized applications.

Thus, the *Code* can be thought of as the "horizontal" standard—or governing regulatory document—for electrical safety, with general wiring rules that apply broadly across all types of installations. The other publications can be thought of as "vertical" standards with specialized wiring rules that apply only to a particular type of installation (**Figure 6.2**). Although the rules in other NFPA standards can increase or add to minimum *Code* requirements, they can't establish less stringent electrical rules for specialized applications. Other NFPA standards can't set *lower* electrical safety requirements.

Vertical electrical standards supplement and extend the general wiring rules of the *NEC*. Rather than repeat all of the wiring rules needed for fire alarms, emergency power systems, or other applications, they use the *NEC* as a general source for most wiring methods. They contain only additional rules that don't appear in the *Code*.

Most of these vertical standards are published by NFPA and referenced in FPNs. They can be thought of as interlocking requirements and engineers must use all of them when designing electrical systems for buildings and similar structures. These vertical NFPA standards are listed here, in the order of their numerical designations:

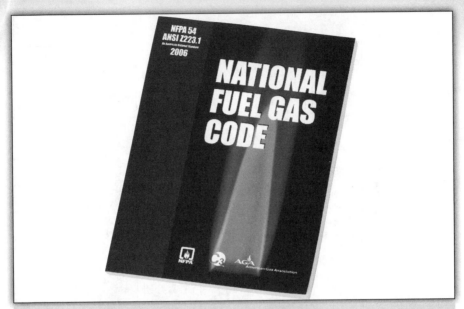

Figure 6.3 Cover of NFPA 54, *National Fuel Gas Code.* © 2006 NFPA.

NFPA 54, *National Fuel Gas Code*

NFPA 54 is adopted for regulatory use by states, cities, and counties (**Figure 6.3**). It covers installation of fuel gas piping systems, gas utilization equipment, and related accessories.

NFPA 54 has several electrical rules that supplement those of the *National Electrical Code.* The requirements listed here are paraphrased and/or summarized, rather than reproduced verbatim:

- **6.13.2** specifies that gas piping shall not be used as a grounding conductor or electrode. *NEC* Section 250.52(B)(1) also prohibits this.
- **6.13** requires that electrical circuits not use gas piping or components as conductors, except for low-voltage control circuits, ignition circuits, and electronic flame detection devices operating at 50 volts or less.
- **6.15.1** states that electrical connections to control devices shall conform to the *National Electrical Code.*
- **6.15.2** states that electrical safety controls shall fail safe (shut off the flow of gas) when power is interrupted.
- **8.6.1** requires that electrical connections to gas utilization equipment conform to the *National Electrical Code.*
- **8.6.2** specifies that electrical ignition and control devices shall not permit unsafe operation of gas utilization equipment when power is interrupted or restored.
- **8.6.4** states that central heating equipment shall be supplied from an individual branch circuit. All gas utilization equipment with electrical controls shall be supplied by a permanently live electrical circuit (one not controlled by a light switch).

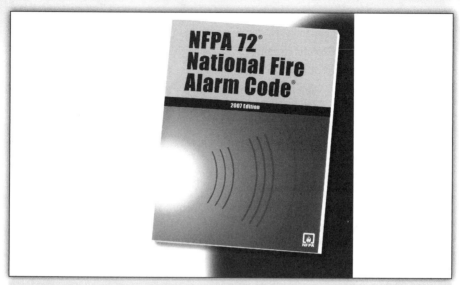

Figure 6.4 Cover of NFPA 72, *National Fire Alarm Code*. © 2007 NFPA.

NFPA 72, *National Fire Alarm Code*

NFPA 72 is adopted for regulatory use by states, cities, and counties. It covers all aspects of fire alarm system design and equipment performance (**Figure 6.4**). The *NEC* contains wiring rules for fire alarm systems, and refers to NFPA 72 in the FPN to 760.1.

However, NFPA 72 also includes a number of additional electrical rules that supplement those of the *National Electrical Code*. The requirements listed here are paraphrased and/or summarized, rather than reproduced verbatim:

- **4.4.1.3 Power Supply Sources** states that fire alarm systems shall be supplied by two independent power supplies, one primary and one secondary.

- **4.4.1.4 Primary Power Supply** states that fire alarm systems shall be supplied by a "dedicated branch circuit." (This term isn't defined in NFPA 70 or NFPA 72, but can be assumed to mean "individual branch circuit.")

- **4.4.1.4.2.2** states that the circuit disconnecting means shall have a red marking and be identified as "fire alarm circuit."

- **4.4.1.5 Secondary Power Supply** states that the secondary power supply shall consist of a storage battery or dedicated branch circuit supplied by a generator.

- **5.5 Requirements for Smoke and Heat Detectors** specifies required locations for these devices.

- **5.12.4** specifies mounting heights of manual fire alarm boxes.

- **5.12.6, 5.12.7, and 5.12.8** specify required locations of manual fire alarm boxes.

- **6.4.2.2.2** states that Class A circuits shall be installed so that the outgoing and return conductors are routed separately, and do not share the same cable, raceway, or enclosure.

- **7.4.6 Location of Audible Notification Appliances** specifies mounting heights of audible notification appliances.
- **7.5.4.1 Spacing in Rooms** specifies locations of visible appliances.
- **7.5.4.2 Spacing in Corridors** specifies locations of visible appliances.
- **9.4 Alarm Transmission Equipment (Auxiliary Boxes, Master Boxes, and Street Boxes)**
  - **9.4.2.1.11.2** states that exterior wire shall be installed in conduit or EMT per Chapter 3 of the *NEC*.
- **9.7 Public Cable Plant**
  - **9.7.1.1.2** states that circuit to a public box installed inside a building shall be installed in RMC, IMC, or EMT.
  - **9.7.1.2.3** requires that interior cables shall comply with the *NEC*.
  - **9.7.1.2.3** requires that signaling wires that may introduce a hazard shall be protected in accordance with the *NEC*.
  - **9.7.1.5.2** states that leads to boxes on poles shall have 600-volt insulation approved for wet locations as defined in the *NEC*.
  - **9.7.1.6.2** specifies that conductors inside building terminals shall be installed in nonflexible raceways, per Chapter 3 of the *NEC*.
  - **9.7.1.6.4.2** states that wire terminals, terminal boxes, splices, and joints inside buildings shall comply with the *NEC*.
- **11.5.1.1 Smoke Detection** specifies required locations of smoke alarms in one- and two-family dwelling units.
- **11.6.1 Smoke and Heat Alarms** requires that alarms in one- and two-family dwelling units be powered by a commercial light and power source with rechargeable battery backup.
- **11.8.3 Smoke Detectors and Smoke Alarms** specifies locations in one- and two-family dwelling units.
- **11.8.4 Heat Detectors and Heat Alarms** specifies locations in one- and two-family dwelling units.

NFPA 75, *Protection of Information Technology Equipment*

NFPA 75 isn't typically adopted for regulatory use by states, cities, and counties. Instead, it is referenced in customer construction guidelines and RFPs. NFPA 75 describes protection of information technology equipment and rooms from damage by fire or its associated effects: smoke, heat, corrosion, and water.

The *NEC* contains most power supply and wiring rules for information technology equipment, and refers to NFPA 75 in the FPN to 645.1. However, NFPA 75 also includes a few additional electrical rules that supplement those of the *National Electrical Code*:

- **8-3** covers electrical service to panelboards serving information technology rooms.
- **8-4** covers supply circuits and interconnecting cables contained within information technology rooms.

NFPA 79, *Electrical Standard for Industrial Machinery*

NFPA 79 typically isn't adopted for regulatory use by states, cities, and counties. Instead, it is referenced in customer construction guidelines. NFPA 79 describes construction and operation of industrial control equipment.

The *NEC* contains most rules for overcurrent protection and supply conductors to industrial machines, and refers to NFPA 79 in the FPN to 670.1.

NFPA 79 primarily deals with design, construction, safety requirements, mounting, and locations of electrically powered industrial machines, along with connections between different components of industrial manufacturing systems. However, it also includes a number of electrical rules that supplement those of the *National Electrical Code*, many of them based on harmonization with IEC standards:

- **5.3 Supply Circuit Disconnecting (Isolating) Means** specifies types, locations, and marking requirements for disconnecting means of industrial machines.

- **5.5 Devices for Disconnecting (Isolating) Electrical Equipment** provides performance requirements for devices that disconnect (isolate) industrial machines to permit work to be performed in a deenergized condition.

- **6.2.3 Enclosure Interlocking** describes methods of interlocking enclosures to prevent contact with energized parts (and permits methods for qualified persons to defeat such interlocks).

- **7.2.1.4** provides special overcurrent protection rules for slash-rated systems such as 120/240 or 277/480 volts.

- **7.9 Power Factor Correction Capacitors** defines performance requirements for capacitors installed for motor power factor correction.

- **8.2 Equipment Grounding (Protective Bonding) Circuit** describes a special type of grounding (bonding) circuit for industrial machines.

- **9.1.2.1 AC Control Circuit Voltages** states that, with certain exceptions, control circuit voltage shall not exceed 120 volts AC or 250 volts DC.

- **Chapter 12—Conductors, Cables, and Flexible Cords** contains numerous special rules based on the fact that wiring associated with industrial machines is often subject to flexing, is shielded due to electromagnetic compatibility (EMC) concerns, and has different ampacities than those in the *NEC* for various reasons.

- **Chapter 13—Wiring Practices** includes numerous special rules, including conductor color coding, connectors molded onto the ends of conductors, and requirements for raceways.

- **Chapter 15—Accessories and Lighting** specifies safety requirements for receptacles and lighting supplied as an integral part of industrial machinery

- **16.4 Machine Nameplate Data** specifies required information.

- **Chapter 17—Technical Documentation** requires detailed installation and circuit diagrams.

- **Chapter 19—Servo Drives and Motors** specifies overload protection requirements for servo motors without such protection integrally installed.

NFPA 99, *Standard for Health Care Facilities*

NFPA 99 is adopted for regulatory use by most states, cities, and counties. Its purpose is to minimize fire, explosion, and electricity hazards in health care facilities such as hospitals, clinics, ambulatory care centers (often known as urgent care centers), dental offices, nursing homes, and limited care facilities (including drug treatment centers and adult day care centers). NFPA 99 doesn't apply to veterinary facilities.

The *NEC* contains most wiring rules for health care facilities. However, NFPA 99 contains a number of additional electrical rules that supplement those of the *National Electrical Code*. Some but not all of these rules are extracted for use in the *NEC*, as described in the FPN at the beginning of Article 517.

NFPA 99 requirements listed here are paraphrased and/or summarized, rather than reproduced verbatim. Different types of health care facilities have different levels of electrical safety requirements based on such factors as the degree of patient incapacitation, the level of care delivered, and whether oxygen and flammable inhalants (anesthetics) are used.

- **Chapter 4—Electrical Systems** covers performance, maintenance, and testing of electrical systems (both normal and essential) used within health care facilities generally.
- **Chapter 8—Electrical Equipment** covers performance, maintenance, and testing of electrical equipment used within health care facilities generally.
- **Chapter 10—Manufacturer Requirements** applies to equipment manufactured for use in the delivery of health care. Chapter 10 of NFPA 99 provides a basis for equipment listing, but isn't of direct interest to engineers designing electrical systems for health care facilities.
- **13.4.1.2.6** identifies specific requirements for electrical systems in hospitals.
- **14.3.4** specifies requirements for electrical systems in ambulatory care centers.
- **17.3.4** gives specific requirements for electrical systems in nursing homes.
- **18.3.4** identifies specific requirements for electrical systems in limited care facilities.
- **20.2.7** specifies requirements for electrical systems in hyperbaric facilities.
- **21.3.4** specifies requirements for electrical systems in freestanding birthing centers.
- **Annex D** covers principles of design and use for high-frequency electrical and electronic appliances used in health care facilities.

NFPA 101, *Life Safety Code*

NFPA 101 deals with classification of occupancies and establishes requirements for emergency egress pathways, including required location of exit signs and lighting (**Figure 6.5**). *NEC* Article 700 specifies wiring requirements for exit signs and emergency lighting, including construction and performance requirements for unit equipment (sometimes called "lunch boxes" in the field). NFPA 101 rules that supplement those of the *NEC* are paraphrased and/or summarized here, rather than reproduced verbatim:

- **7.8 Illumination of Means of Egress** defines general requirements and illumination levels for illuminating exit pathways from buildings.
- **7.9 Emergency Lighting** provides general requirements for power supplies, testing, product listing, and such. These are supplemented by more exact requirements in Chapters 11–42 on specific occupancies.

Figure 6.5 Cover of NFPA 101, *Life Safety Code.* © 2006 NFPA.

- **7.9.5 Illumination of Signs** provides requirements for illumination of exit signs.
- **9.1.2 Electrical Systems** requires that electrical wiring and equipment comply with the *NEC*.
- **9.6 Fire Detection, Alarm, and Communications Systems** requires that these comply with the *NEC* and NFPA 72.

NFPA 731, *Standard for the Installation of Electronic Premises Security Systems*

NFPA 731 is a performance standard for electronic burglar and security systems of all types (**Figure 6.6**). As a relatively new standard, first published in 2006, it isn't yet widely adopted for regulatory use or referenced in customer construction guidelines. Three *NEC* articles cover wiring used with security systems:

- Article 725—Class 2 and Class 3 wiring
- Article 800—Telephone-type conductors, including category-rated UTP cabling
- Article 820—Coaxial cables

NFPA 731 rules that supplement those of the *NEC* are paraphrased and/or summarized here, rather than reproduced verbatim:

- **4.2.2, 4.2.3 Power Sources** state that premises security systems shall have two independent power supplies, one primary and one secondary. These power supplies shall comply with NFPA 70.
- **4.2.4.1** specifies that the primary power supply shall be a "dedicated branch circuit." (This term isn't defined in NFPA 70 or NFPA 731, but can be assumed to mean "individual branch circuit.")

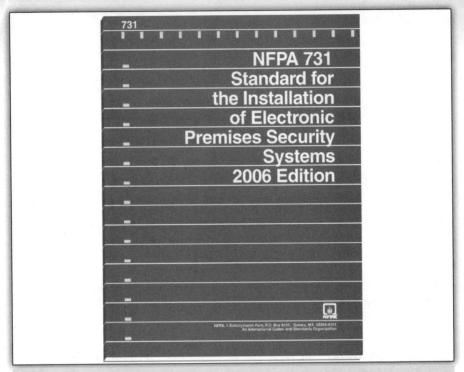

Figure 6.6 Cover of NFPA 731, *Standard for the Installation of Electronic Premises Security Systems.* © 2006 NFPA.

- **4.2.4.2.1** specifies that the dedicated circuit disconnecting means shall have a blue marking and be labeled "premises security circuit."
- **4.2.8.3** states that storage batteries shall comply with *NEC* Article 480.
- **4.2.91** requires that engine generator installations comply with NEC Article 700 and NFPA 110, *Standard for Emergency and Standby Power Systems*. (NFPA 110 is a performance standard and doesn't include electrical wiring rules.)
- **4.5.7** specifies that circuits and equipment shall be protected against transients in accordance with *NEC* Article 800. (This refers to primary protector.)
- **4.5.8** requires that wiring and cabling be installed in accordance with *NEC* Articles 725, 770, and 800. Detailed performance requirements include a mandatory 50 mm (2 in.) separation from light and power conductors, unless either circuit is in a metal raceway.
- **7.7.2.2 Coaxial Cable Distances** lists recommended maximum lengths for different grades of coaxial cable.
- **7.7.2.3 Coaxial Connections** states that all connections shall be made using BNC connectors.

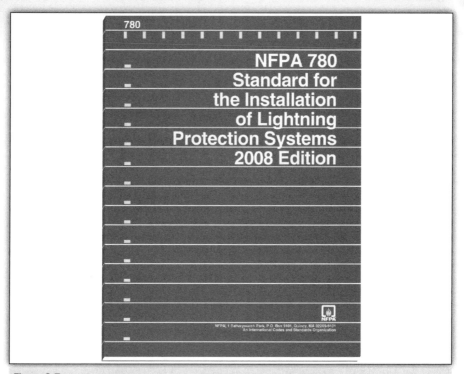

Figure 6.7 Cover of NFPA 780, *Standard for the Installation of Lightning Protection Systems.* © 2008 NFPA.

NFPA 780, *Standard for the Installation of Lightning Protection Systems*

NFPA 780 is adopted for regulatory use by states, cities, and counties (**Figure 6.7**). It is a self-contained standard that covers all design, performance, and installation aspects for lightning protection systems.

NFPA 780 defines the design and installation of lightning protection systems. The *NEC* refers to NFPA 780 in the FPN to 250.106 as indicated here, and contains several other requirements related to lightning protection:

- **250.60 Use of Air Terminals** states that air terminals cannot be used as substitutes for other required grounding electrodes.

- **250.106 Lightning Protection Systems** requires that lightning protection system ground terminals be bonded to the building or structure grounding electrode system. The two FPNs to 250.106 refer to NFPA 780 for more information about lightning protection, and mention two "rules of thumb" for spacing between lightning protection conductors and noncurrent-carrying metal parts of electric equipment.

- **Article 280—Surge Arresters.** The *NEC* doesn't require installation of surge arresters to protect premises wiring systems from lightning damage. This article provides rules for when they are installed.

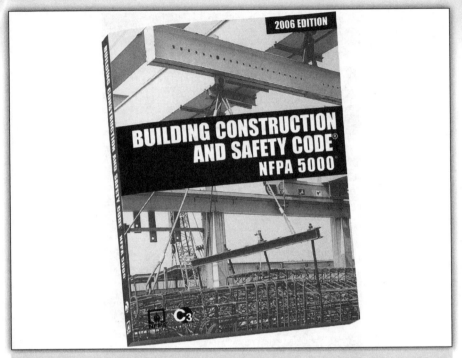

Figure 6.8 Cover of NFPA 5000, *Building Construction and Safety Code.* © 2006 NFPA.

- **800.53 Lightning Conductors** recommends a separation of at least 1.8 m (6 ft) between communications wires and cable on buildings (i.e., exterior conductors) and lightning conductors.

- **800.90 Protective Devices** requires listed primary protectors to be used on communications wiring where exposed to lightning.

- **820.44(F)(3)** recommends a separation of at least 1.8 m (6 ft) between coaxial cable outside buildings and lightning conductors.

- **830.44(I)(3)** recommends a separation of at least 1.8 m (6 ft) between network-powered broadband communications cable on buildings (i.e., exterior conductors) and lightning conductors.

NFPA 5000, *Building Construction and Safety Code*

NFPA 5000 is adopted for regulatory use by states, cities, and counties (**Figure 6.8**). Rules that supplement those of the *NEC* and NFPA 72 are paraphrased and/or summarized here, rather than reproduced verbatim:

- **7.2.3.2.16 Plenum Material Combustibility** defines types of electrical wires and cables, and optical-fiber and communications raceways, permitted in plenums used for environmental air.

- **7.2.3.2.18 Plenum Light Diffusers** requires that plastic diffusers used in air-handling light fixtures be listed and marked for the application.

- **8.7.4** requires that fire detection devices used to close fire doors conform to NFPA 72.

Figure 6.9 Cover of *NFPA 70E, Standard for Electrical Safety in the Workplace.* © 2004 NFPA.

- **8.8.2 Firestop Systems and Devices Required** requires that penetrations for cables, cable trays, conduits, and similar electrical items that pass through fire barriers be protected by a firestop system or device.

- **11.8 Illumination of Means of Egress** establishes requirements for emergency lighting exit signs, including illuminations levels and performance requirements for the emergency power supply.

- **Chapter 51—Energy Efficiency** requires that most buildings meet the requirements of ASHRAE/IESNA 90.1, *Energy Standard for Buildings Except Low-Rise Residential Buildings.* It also requires one- and two-family dwellings to comply with ASHRAE/IESNA 90.2, *Energy Efficient Design of Low-Rise Residential Buildings.*

- **Chapter 52—Electrical Systems** requires that all electrical systems and equipment comply with NFPA 70, *National Electrical Code.* It requires that in new or substantially restored buildings within flood hazard areas, electrical systems and equipment below the design flood elevation shall meet the requirements of SEI/ASCE 24, *Flood Resistant Design and Construction.*

- **Chapter 52—Fire Protection Systems and Equipment** provides detailed performance requirements for fire detection, alarm, and communication systems.

NFPA 70E, *Standard for Electrical Safety in the Workplace*

NFPA 70E covers safe work practices for installing and maintaining electrical systems (**Figure 6.9**). It requires the use of techniques such as lockout/tagout, flame resistant (FR) clothing, and voltage-rated (V-rated) tools to protect workers from electrical shock and arc-flash hazards.

Although compliance with *NFPA 70E* rules isn't required to design and specify electrical systems in buildings and similar structures, every electrical engineer should have a basic familiarity with this important industry safety standard. Plant/facility engineers who supervise or contract for electrical construction and maintenance work may need detailed knowledge of how *NFPA 70E* safety rules are applied. Training is available through the NFPA and other vendors.

Relationship of NFPA 70E to the *National Electrical Code*

NFPA 70E has essentially the same scope as the *NEC* and includes many of the same definitions, but the two documents have different purposes. The *National Electrical Code* applies to electrical installations, while *NFPA 70E* applies to workplaces, including construction sites and places where electrical maintenance work is performed. In other words, the *NEC* describes how to install electrical systems, while *NFPA 70E* describes how to perform the work safely.

***NEC* application and enforcement.** The *National Electrical Code* covers design and construction of electrical systems in buildings and similar structures. Compliance with the *NEC* is required by state and municipal ordinances, and enforced by electrical inspectors working for state and local building departments. However, the *NEC* doesn't have rules governing the procedures by which electrical construction and maintenance work are performed.

NFPA 70E application and enforcement. *NFPA 70E, Standard for Electrical Safety in the Workplace*, covers safe work practices for installing and maintaining electrical systems in buildings and similar structures. The safety standard isn't adopted by states and municipalities for regulatory use, which means state and municipal electrical inspectors don't enforce *NFPA 70E*.

Instead, construction customers often require that electrical contractors comply with *NFPA 70E* while working on their facilities. Increasingly, these customers require that electricians and contractors working for them provide evidence that they've been trained in *NFPA 70E* safe work practices. Doing so reduces the customers' liability exposure and insurance premiums.

As previously explained, the *National Electrical Code* is a self-contained regulatory document that doesn't include mandatory references to other industry codes and standards. *NFPA 70E* is mentioned only in the *NEC* Article 100 definition of *Qualified Person*:

Qualified Person. One who has skills and knowledge related to the construction and operation of the electrical equipment and installations and has received safety training to recognize and avoid the hazards involved.

FPN: Refer to NFPA 70E-2004, *Standard for Electrical Safety in the Workplace*, for electrical safety training requirements.

Relationship Between NFPA 70E and OSHA Regulations

NFPA 70E is a private-sector, voluntary standard that parallels the following OSHA safety regulations:

- OSHA 29 CFR 1910, General Industry Standards, Subpart S—Electrical (covers maintenance and repairs on existing systems)
- OSHA 29 CFR 1926, Construction Industry Standards, Subpart K—Electrical (these cover new construction)

Most electrical contractors are required to follow both sets of OSHA safety regulations, depending on the type of work they are performing: 1926 covers new construction, while CFR 1910 covers maintenance. The two standards are similar, but not quite matching, which often causes confusion.

OSHA involvement. *NFPA 70E* originally was developed at the request of OSHA. Because the federal government rulemaking process is slow and cumbersome, it's hard to keep OSHA regulations up-to-date with current technology and work practices. For example, the 1926, Subpart K regulations still reference the 1984 *National Electrical Code.*

For this reason, private industry uses the *NFPA 70E* standard to "lead" the two sets of OSHA workplace safety regulations that govern electrical work: 1910 Subpart S for general industry applications and 1926 Subpart K for the construction industry. OSHA personnel participate in the *NFPA 70E* committee.

NFPA 70E enforcement. The practical result of complying with the safe work practices defined in *NFPA 70E* is to also comply with the applicable OSHA regulations. Although OSHA safety and health compliance officers do not enforce *NFPA 70E per se*, there is a growing tendency for them to rely on *NFPA 70E* under the so-called "general duty" clause, which requires employers to maintain safe workplaces for their employees and allows OSHA inspectors to consider prevailing industry standards.

Summary of Contents

Chapter 1 of *NFPA 70E* defines safety-related work practices applicable to all electrical construction and maintenance work, including the following:

Establishing an Electrically Safe Work Condition [Article 120]. *NFPA 70E* recommends that all electrical work be performed with the power turned off, on deenergized systems. It permits three methods of controlling hazardous electrical energy to prevent injury to employees:

1. Individual employee control
2. Simple lockout/tagout
3. Complex lockout/tagout

Working On or Near Live Parts [Article 130]. When it isn't feasible to work deenergized (for example, in process industries), *NFPA 70E:*

- Requires a "live work permit" signed by the customer stating the justification for working on energized conductors and equipment.
- Provides a method for calculating the available incident energy under arc-flash conditions.

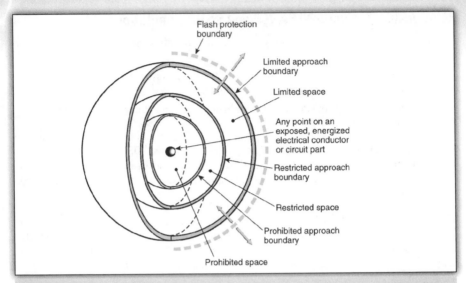

Figure 6.10 Limits of approach. Reprinted with permission from NFPA 70E®-2004, *Electrical Safety in the Workplace,* Copyright © 2004, National Fire Protection Association, Quincy, MA. This reprinted material is not the complete and official position of the NFPA on the referenced subject, which is represented only by the standard in its entirety.

- Defines a flash protection boundary and safe approach distances under various circumstances (**Figure 6.10**).
- Defines different levels of flame-resistant (FR) clothing and PPE that employees use, based on the incident energy exposure. These levels are called *Hazard/Risk Category Classifications.*
- Requires alerting techniques such as safety signs, barricades, and attendants to protect nonqualified persons from electrical hazards.
- Defines other safe work practices (for example, don't wear conductive jewelry and metal-frame glasses, and don't reach blindly into areas that might contain exposed live parts).

Hazard/risk categories. The heart of the *NFPA 70E* standard is two tables that define required levels of PPE for a number of typical electrical tasks (for example, working on a motor control center rated up to 600 volts):

- **Table 130.7(C)(9)(a)** defines hazard/risk category classifications ranging from one to four for each typical electrical task, and states whether V-rated gloves and V-rated tools are needed.
- **Table 130.7(C)(10)** defines the types of FR clothing and other PPE needed for each Hazard/Risk Category.

Other requirements. Chapters 2, 3, and 4 of *NFPA 70E* define more specific safety-related techniques for equipment maintenance, equipment installation, and special equipment (such as batteries, lasers, and power electronics). Informational annexes cover such subjects as limits of approach, calculating arc-flash incident energy, a sample lockout/tagout procedure, a job briefing and planning checklist, and a simplified two-category FR system that covers most common electrical work tasks.

International Code Council (ICC) Publications

The ICC publishes nonelectrical building codes that aren't developed under ANSI-accredited procedures. Two ICC documents contain electrical requirements that engineers should be aware of when designing and specifying projects in those parts of the country where these codes are adopted.

International Building Code (IBC)

Section 2701—General requires that electrical components, equipment, and systems comply with the ICC Electrical Code, which is included in Appendix K of the IBC. The ICC Electrical Code states that, in general, electrical systems and equipment shall be installed in accordance with NFPA 70 and includes a number of additional rules that differ from those of the *National Electrical Code*. For example, the IBC has no limitations on the uses of nonmetallic-sheathed cable (Type NM and NMC) based on building construction type, as the *NEC* does in Article 334.

Section 2702—Emergency and Standby Power Systems specifies a number of occupancies for which emergency and standby power systems are required; states that they shall be installed in accordance with the *ICC Electrical Code*, NFPA 110, and NFPA 111; and states that these emergency and standby power systems shall be maintained and tested in accordance with the *International Fire Code*.

International Residential Code

The International Residential Code (IRC) is a self-contained building code for one- and two-family dwellings, which contains a number of electrical rules:

- Chapter 34—Electrical Definitions
- Chapter 35—Services
- Chapter 36—Branch Circuit and Feeder Requirements
- Chapter 37—Wiring Methods
- Chapter 38—Power and Lighting Distribution
- Chapter 39—Devices and Luminaires
- Chapter 40—Appliance Installation
- Chapter 41—Swimming Pools
- Chapter 42—Class 2 Remote-Control, Signaling, and Power-Limited Circuits

In general, the technical content of IRC electrical rules is similar to that of *NEC* rules for one- and two-family dwellings, but they are organized and numbered differently.

Conclusion

The *National Electrical Code* is the national building code for electrical installations. However, many other industry codes and standards have requirements that supplement those of the *NEC* for specialized applications. This unit characterizes the *Code* as a horizontal safety standard with a broad scope, augmented by a family of vertical standards with narrower scopes. It summarizes the "interlocking" requirements in other codes and standards that supplement *NEC* wiring rules.

UNIT 7

Making the *NEC*

Introduction

All ANSI-approved standards are revised at intervals, typically every 5 years, but because it is the U.S. national wiring rules, the *National Electrical Code* is revised more often. To stay up-to-date with current electrical technology and construction methods, the *NEC* is updated every 3 years; the revision process itself takes nearly 2 years to complete.

There are also formal procedures for issuing formal interpretations and tentative interim amendments between regular editions of the *NEC*. This unit covers the following:

- National Electrical Code Committee
- *NEC* Revision Process
- Formal Interpretations and Tentative Interim Amendments
- NFPA Policies and Procedures
- Effective Participation in the *NEC* Process
- Writing Effective Code Proposals and Public Comments

National Electrical Code Committee

Committee Structure

The National Electrical Code is written by 20 different technical subcommittees known as *Code-Making Panels (CMPs)*. Rosters of each CMP appear at the beginning of the *Code* book, along with the articles for which they are responsible. CMP scopes can be summarized as follows:

- CMP-1—General rules, definitions
- CMP-2—Branch circuits
- CMP-3—Wiring methods, including low-voltage control and fire alarm wiring
- CMP-4—Services, feeders
- CMP-5—Grounding and bonding, surge arresters
- CMP-6—Conductors, flexible cords and cables
- CMP-7—Cables
- CMP-8—Raceways, conduits, tubing
- CMP-9—Power distribution equipment
- CMP-10—Overcurrent protection
- CMP-11—Motors and motor control, heating and air conditioning
- CMP-12—Cranes, hoists, and lifts; welders, data processing rooms
- CMP-13—Power systems including transformers, emergency and standby power systems, solar photovoltaics, and fuel cells
- CMP-14—Hazardous (classified) locations
- CMP-15—Special occupancies including hospitals, places of assembly, carnivals, and fairs

- CMP-16—Communications systems
- CMP-17—Pools, heating equipment, appliances
- CMP-18—Lighting, electric signs, wiring devices
- CMP-19—Manufactured buildings, mobile homes, trailer parks, marinas, agricultural buildings
- CMP-20—Critical operations power systems (COPS)

The National Electrical Code Technical Correlating Committee (NEC-TCC) supervises the work of the 20 CMPs and manages the overall *Code* process. The NEC-TCC determines policy, approves appointments to CMPs, and reviews the actions taken by all CMPs to ensure consistency throughout the *National Electrical Code*.

Code Panel Membership

CMP members represent organizations, rather than serving as individuals. Each organization can have a principal and alternate representative. The following electrical industry organizations have members on all or most CMPs:

- American Chemistry Council (ACC)
- Edison Electric Institute (EEI)
- Independent Electrical Contractors (IEC)
- International Association of Electrical Inspectors (IAEI)
- International Brotherhood of Electrical Workers (IBEW)
- Institute of Electrical and Electronics Engineers (IEEE)
- National Electrical Contractors Association (NECA)
- National Electrical Manufacturers Association (NEMA)
- Underwriters Laboratories, Inc. (UL)

Other industry groups with a more limited interest in the *NEC* have members only on selected panels. Examples include the following:

- American Society for Healthcare Engineering, CMP-15
- Crane Manufacturers Association of America, CMP-12
- International Sign Association, CMP-18
- National Spa and Pool Institute, CMP-17
- Recreational Vehicle Industry Association, CMP-19
- Society of the Plastics Industry, CMP-8

NEC Revision Process

Revising the *National Electrical Code* is a two-stage process that gives all interests a chance to participate and have their views considered. The *Code* revision process can be summarized as follows.

ROP (Proposal) Stage

Proposals. Typically, between 3,500 and 4,000 proposals are submitted every 3 years to revise the *NEC*. They range from minor editorial suggestions to revised sections to entirely new articles. Any person can submit a change proposal and each is considered by the appropriate CMP; there is no prescreening process.

First panel meetings. All 20 CMPs meet to review the change proposals submitted for their parts of the *Code*, voting on each one.

ROP issued. These actions, along with the CMP reasons for their actions, are published for public review in a document called the "Report on Proposals" (ROP). It is available free of charge in three formats: online, on CD, and as a paperbound book (**Figure 7.1**).

2007 Annual Revision Cycle

National Electrical Code ®
Committee Report

This Report contains the proposed amendments for the 2008 *National Electrical Code*® for public review and comment prior to October 20, 2006, and for consideration at the NFPA June 2007 Association Technical Meeting

NOTE: The proposals contained in this NEC Report on Proposals (ROP) and the comments addressed in a follow-up Report on Comments (ROC) will be presented for action at the NFPA June 2007 Association Technical Meeting to be held June 3–7 in Boston, MA, only when proper Amending Motions have been submitted to the NFPA by the deadline of May 4, 2007. For more information on the new rules and for up-to-date information on schedules and deadlines for processing NFPA Documents, check the NFPA website (www.nfpa.org) or contact NFPA Standards Administration.

National Fire Protection Association
NFPA® 1 BATTERYMARCH PARK, QUINCY, MA 02169-7471

Figure 7.1 The *NEC* report on proposals describes actions taken on change proposals by CMPs.

ROC (Comment) Stage

Public comments. In the second stage of the *Code* revision process, anyone may submit public comments. Frequently, these comments ask CMPs to reconsider their original actions on proposals, based on additional technical information or for other reasons.

Second panel meetings. The 20 CMPs meet a second time to discuss and vote on public comments.

ROC issued. CMP actions are published for public review in a second free publication called the *Report on Comments (ROC)*.

Final Approval

The new *National Electrical Code* (along with other NFPA codes and standards) is voted on by the association membership at NFPA's annual meeting, called the World Fire Safety Conference & Exposition. Typically, there are a number of last-minute amending motions. The NFPA Standards Council then approves and issues the *NEC*.

NEC: An American National Standard

NFPA's codes and standards development procedures are accredited by the American National Standards Institute (ANSI), the Washington, DC–based federation of U.S. standards-developing organizations. Thus, the *National Electrical Code* is an ANSI standard. The terms "ANSI standard" and "American National Standard" are synonymous.

How to Submit *Code* Proposals

Each *NEC* book contains a proposal submission form and revision schedule for the next edition, including the deadline for submitting proposals to NFPA (**Figure 7.2**).

Formal Interpretations and Tentative Interim Amendments

Formal Interpretations

The meaning of the *National Electrical Code* is completely represented by its text and figures. No NFPA staff person or member has authority to officially interpret these rules. Any informal interpretation or opinion expressed is just that; an informal, personal, opinion.

However, there is a procedure for requesting a formal interpretation of *Code* rules. Any user can request a formal interpretation by submitting a question worded in such a way that it can be answered with a single word: Yes or No. This question then is voted on by the CMP responsible for that article.

Tentative Interim Amendments

The *NEC* is revised at shorter intervals than other industry codes and standards. Still, sometimes the need arises to make a partial revision between regularly scheduled editions. In the NFPA system, these are called tentative interim amendments (TIAs).

FORM FOR PROPOSALS FOR 2008 *NATIONAL ELECTRICAL CODE*®

Mail to: Secretary, Standards Council
           National Fire Protection Association
           1 Batterymarch Park, P.O. Box 9101
           Quincy, Massachusetts 02169-7471

| FOR OFFICE USE ONLY |
| Log # _____ |
| Date Rec'd _____ |

Fax to: (617) 770-3500

Notes: 1. All proposals must be received by 5:00 p.m. EST on Friday, November 4, 2005. Proposals received
              after 5:00 p.m. EST, Friday, November 4, 2005, will be returned to the submitter.
           2. Type or print legibly in black ink. Limit each proposal to a SINGLE section. Use a separate copy for
              each proposal.
           3. If supplementary material (photographs, diagrams, reports, etc.) is included, you may be required to
              submit sufficient copies for all members and alternates of the technical committee.

Please indicate in which format you wish to receive your ROP/ROC: ❏ electronic   ❏ paper   ❏ download

Date _____   Name _____   Tel. No.: _____

Company _____

Street Address _____

Organization Represented (if any) _____

1. Section/Paragraph _____

2. Proposal Recommends (check one)   ❏ new text   ❏ revised text   ❏ deleted text

3. Proposal (include proposed new or revised wording or identify wording to be deleted). Note: Proposed text
   should be in a legislative format: i.e., use underscore to denote wording to be inserted (<u>inserted wording</u>) and
   strike-through to denote wording to be deleted (~~deleted wording~~).

4. Statement of Problem and Substantiation for Proposal. Note: State the problem that will be resolved by your
   recommendation; give the specific reason for your proposal and include copies of the tests, research papers,
   fire experience, etc. If more than 200 words, it may be abstracted for publication.

5. ❏ This Proposal is original material. Note: Original material is considered to be the submitter's own idea based
   on or as a result of his/her own experience, thought, or research and, to the best of his/her knowledge, is not
   copied from another source.

   ❏ This Proposal is not original material; its source (if known) is as follows: _____

| If you need further information on the standards-making process, please contact the
Standards Administration Department at (617) 984-7249.
For technical assistance, please call NFPA at (617) 770-3000. |

*I hereby grant the NFPA all and full rights in copyright, in this proposal, and I understand that I acquire no rights
in any publication of NFPA in which this proposal in this or another similar or analogous form is used.*

_____
Signature (required)

PLEASE USE SEPARATE FORM FOR EACH PROPOSAL

Figure 7.2 A proposal submission form for the next edition is printed in the *National Electrical Code*.

As with formal interpretations, a proposed TIA (adding new language, revising existing language, or deleting language) is voted on by the CMP responsible for that article. Usually, TIAs are approved only when there is a strong safety justification that persuades the CMP to take action before the next regular *Code* revision.

When a TIA is approved, it is included in subsequent printings of that edition of the *NEC*. The amendment is called "tentative" because it hasn't been processed through the entire NFPA standards-making procedure and "interim" because it is effective only between editions of NFPA 70.

A TIA automatically becomes a change proposal for the next edition of the *Code*, at which time it goes through the normal two-stage revision process.

Limitations of TIAs. When TIAs are issued for the *National Electrical Code* or other NFPA documents, it's difficult to publicize them effectively throughout the affected industries. Early printings of the *NEC* or other codes and standards, already in thousands of users' hands, don't include TIA text. Also, when states and municipalities adopt the *NEC* and other NFPA documents for regulatory enforcement, they adopt the version in effect at that particular time. TIAs issued later typically aren't enforced by state and local building departments.

NFPA Policies and Procedures

The *National Electrical Code*, and all other NFPA codes and standards, are developed through ANSI-accredited consensus procedures that allow for broad public participation at every stage of the development process.

No other voluntary code or standard adopted for regulatory use has such broad public participation, with thousands of change proposals and comments. No other standards-developing organization has technical committees (i.e., CMPs) that debate change proposals and public comments in an atmosphere of such complete openness. All deliberations of CMPs, the *NEC* Technical Correlating Committee, and NFPA's Standards Council are clearly documented and freely available for public review.

This section summarizes NFPA procedures and editorial style manuals that shape the making of the *Code*, with brief explanations of how they work together to form an organic whole that can be summed up in a single word: *consensus*. The procedures and manuals described here are available at the NFPA Web site (www.nfpa.org; click on *Codes & Standards*).

Regulations Governing Committee Projects

NFPA's Regulations Governing Committee Projects (generally known as the "Regulations") spell out the process of developing and revising all NFPA codes and standards. They define the following:

- Duties of technical committees, technical correlating committees, the standards council, the board of directors, and NFPA staff
- Submitting change proposals and public comments on NFPA documents
- Qualifications and procedures for membership on NFPA committees

- Conducting meetings, voting procedures, filing appeals
- Procedures for TIAs and formal interpretations

Supplemental Operating Procedures for the National Electrical Code Committee

Due to the size of the National Electrical Code Committee and the complexity of the code-making process, there are special *NEC* procedures that supplement the Regulations.

NFPA Manual of Style

The manual of style, which is itself a standard, provides rules for writing (as opposed to developing) NFPA codes and standards. It covers everything from mandatory and permissive language to section numbering to table formats and units of measurement.

National Electrical Code Style Manual

The *Code* is written in a different format and editorial style than other NFPA documents. These differences include its numbering system, which divides chapters into articles, and the use of FPNs. For this reason, there is a separate *National Electrical Code Style Manual* that supplements the *NFPA Manual of Style*.

In addition to purely editorial matters, the *NEC Style Manual* contains a number of rules that shape the overall structure and language of the *Code*. It's intended to be used by both CMPs and NFPA staff as a practical working tool to assist in making the *NEC* as clear, usable, and unambiguous as possible. For this reason, a working knowledge of this manual is helpful to anyone who uses the *National Electrical Code* on a regular basis.

Effective Participation in the *NEC* Process

CMPs consist primarily of organizational representatives, rather than individuals. Most members are sponsored by professional societies and industry groups, such as those listed at the beginning of this unit, to ensure participation by a broad cross-section of the electrical industry.

Different organizations have various processes for appointing representatives to outside technical committees, but many have a codes and standards committee, regulatory affairs board, standards and safety committee, or similar body that governs their participation in industry codes and standards activities. Electrical engineers and others selected as members of CMPs and other industry standards committees are generally those who are active in their industry's professional and trade groups.

Public Proposals and Comments

Others can participate in the *NEC* process by submitting change proposals and public comments as described earlier in this unit. The most effective proposals and comments are those that propose specific wording and are supported by detailed technical substantiations.

Other NFPA Codes and Standards

All NFPA codes and standards are created and maintained using the same ANSI-accredited development procedures as the *NEC*. Each one has a technical committee made up of organizational representatives (large documents such as NFPA 72, NFPA 101, and NFPA 5000 have multiple technical committees similar to CMPs) and is revised in a two-stage process at regular intervals, usually every 4 or 5 years.

Other Industry Codes and Standards

Regulatory documents developed by other organizations, such as the *National Electrical Safety Code* (IEEE) and the *International Building Code* (ICC), have similar development procedures involving technical committees and public input, though the details vary.

Writing Effective Code Proposals and Public Comments

Approximately 3,500–4,000 change proposals are submitted during each *NEC* revision cycle, along with a similar number of public comments. Many end up being rejected by *Code*-making panels. Most proposals and comments are rejected for substantive reasons, because the CMP having jurisdiction believes they aren't feasible or won't improve electrical safety. Some, however, are rejected for procedural reasons, because they don't comply with references such as the NFPA Regulations Governing Committee Projects and the *National Electrical Code Style Manual*.

The most common procedural violations that CMPs cite when rejecting *Code* proposals and public comments are listed here (these are actual excerpts from ROPs and ROCs). Avoiding these basic mistakes when submitting *NEC* proposals and comments improves an engineer's chances of having a submission judged on its technical merits.

All NFPA Codes and Standards

The following procedural objections apply to change proposals and public comments submitted for NFPA codes and standards in general.

Panel Statement: No proposed text was submitted, as required by 4.3.3(c) of the NFPA Regulations Governing Committee Projects.

Panel Statement: No statement of a problem and substantiation was provided, as required by 4.3.3(c) of the NFPA Regulations Governing Committee Projects.

Panel Statement: The section of the document to which the proposal is directed was not identified, as required by 4.3.3(c) of the NFPA Regulations Governing Committee Projects.

Panel Statement: Definitions shall not contain requirements, in accordance with 2.3.2.3 of the *NFPA Manual of Style*.

Panel Statement: The proposal contains a vague or unenforceable term prohibited by 2.2.2.3 of the *NFPA Manual of Style*.

Panel Statement: The word "Listed" is an official NFPA term, whose definition shall not be altered except by the NFPA Standards Council, according to 3.3.6.1 of the Regulations Governing Committee Projects. (Section 3.3.6.1 also contains a list of official NFPA terms.)

Conclusion

This unit describes the procedures by which the *National Electrical Code* is revised at 3-year intervals, the structure of the National Electrical Code Committee, how formal interpretations and TIAs are developed, and how engineers and other *Code* users can participate in these processes.

APPENDIX A

Electrical Design Standards

Introduction

There are many voluntary industry codes and standards that electrical engineers use as design references. With nearly 13,000 ANSI-approved standards in existence, and many other standards of various kinds not developed under consensus procedures, any listing of recommended design guides for electrical engineers is bound to be incomplete. However, most lists of indispensable guides for designing, specifying, and installing electric power systems for buildings would include at least those listed in **Tables A.1–A.4**.

Institute of Electrical and Electronics Engineers

445 Hoes Lane
P.O. Box 1331
Piscataway, NJ 08855-1331
Telephone: (800) 678-4333
Fax: (732) 697-8341
www.ieee.org

Table A.1 IEEE Color Book Series

IEEE 141-1993 (R1999)	Recommended Practice for Electric Power Distribution for Industrial Plants	Red Book
IEEE 142-1991	Recommended Practice for Grounding of Industrial and Commercial Power Systems	Green Book
IEEE 241-1990 (R1997)	Recommended Practice for Electrical Power Systems in Commercial Buildings	Gray Book
IEEE 242-2001	Recommended Practice for Protection and Coordination of Industrial and Commercial Power Systems	Buff Book
IEEE 399-1997	Recommended Practice for Power Systems Analysis	Brown Book
IEEE 446-1995	Recommended Practice for Emergency and Standby Power Systems for Industrial and Commercial Applications	Orange Book
IEEE 493-1997	Recommended Practice for the Design of Reliable Industrial and Commercial Power Systems	Gold Book
IEEE 602-1996	Recommended Practice for Electric Systems in Health Care Facilities	White Book
IEEE 739-1995	Recommended Practice for Energy Conservation and Cost-Effective Planning in Industrial Facilities	Bronze Book
IEEE 902-1998	Guide for Maintenance, Operation, and Safety of Industrial and Commercial Power Systems	Yellow Book
IEEE 1015-2006	Recommended Practice for Applying Low Voltage Circuit Breakers Used in Industrial and Commercial Power Systems	Blue Book
IEEE 1100-2005	Recommended Practice for Powering and Grounding Electronic Equipment	Emerald Book

Illuminating Engineering Society of North America

120 Wall Street, 17th Floor
New York, NY 10005-4001
Telephone: (212) 248-5000
Fax: (212) 248-5017
www.iesna.org

Table A.2 Lighting

IESNA HB-9-00	*IESNA Lighting Handbook, 9th Edition*
IESNA AEDG-1-05	*Advanced Energy Design Guide for Small Office Buildings*
IESNA AEDG-2-06	*Advanced Energy Design Guide for Small Retail Buildings*
IESNA PB-72-02	*Architectural Lighting Design*
IESNA RDW-SET-06	*Roadway Lighting Package*
IESNA RP-1-04	*Standard Practice for Office Lighting*
IESNA RP-22-05	*Recommended Practice for Lighting Hospitals and Health Care Facilities*

Telecommunications Industry Association (TIA)

2500 Wilson Boulevard, Suite 300
Arlington, VA 22201-3834
Telephone: (703) 907-7700
Fax: (703) 907-7727
www.tiaonline.org

Table A.3 Telecommunications

ANSI/TIA/EIA-568-B. 1-04	*Commercial Building Telecommunications Cabling Standard*
ANSI/TIA/EIA-569-B-04	*Commercial Building Standard for Telecommunications Wiring Pathways and Spaces*
ANSI/TIA/EIA-570-B-04	*Residential Telecommunications Cabling Standard*
ANSI/TIA/EIA-606-A-02	*Administration Standard for Commercial Telecommunications Infrastructure*
ANSI/TIA/EIA-607-94	*Commercial Building Grounding and Bonding Requirements for Telecommunications*

National Electrical Contractors Association

3 Bethesda Metro Center, Suite 1100
Bethesda, MD 20814
Telephone: (301) 215-4504
Fax: (301) 215-4500
www.neca-neis.org

Table A.4 National Electrical Installation Standards (NEIS)

NECA 1-2006	*Standard Practices for Good Workmanship in Electrical Construction* (ANSI)
NECA 90-2004	*Recommended Practice for Commissioning Building Electrical Systems* (ANSI)
NECA 100-2006	*Symbols for Electrical Construction Drawings* (ANSI)
NECA 101-2006	*Standard for Installing Steel Conduits (Rigid, IMC, EMT)* (ANSI)
NECA 102-2004	*Standard for Installing Aluminum Rigid Metal Conduit* (ANSI)
NECA/AA 104-2006	*Recommended Practice for Installing Aluminum Building Wire and Cable* (ANSI)
NECA/NEMA 105-2002	*Recommended Practice for Installing Metal Cable Tray Systems* (ANSI)
NECA 111-2003	*Standard for Installing Nonmetallic Raceways (RNC, ENT, LFNC)* (ANSI)
NECA 120-2006	*Standard for Installing Armored Cable (Type AC) and Metal-Class Cable (Type MC)* (ANSI)
NECA 200-2002	*Recommended Practice for Installing and Maintaining Temporary Electrical Power at Construction Sites* (ANSI)
NECA 202-2006	*Recommended Practice for Installing and Maintaining Industrial Heat Tracing Systems* (ANSI)
NECA 230-2003	*Standard for Selecting, Installing, and Maintaining Electric Motors and Motor Controllers* (ANSI)
NECA/FOA 301-2004	*Standard for Installing and Testing Fiber Optic Cables* (ANSI)
NECA 303-2005	*Standard for Installing Closed-Circuit Television Systems* (ANSI)
NECA 305-2001	*Standard for Fire Alarm Systems Job Practices* (ANSI)
NECA 331-2004	*Standard for Building and Service Entrance Grounding and Bonding*
NECA 400-2007	*Standard for Installing and Maintaining Switchboards* (ANSI)
NECA 402-2007	*Standard for Installing and Maintaining Motor Control Centers* (ANSI)
NECA/EGSA 404-2007	*Standard for Installing Generator Sets* (ANSI)
NECA 406-2003	*Recommended Practice for Installing Residential Generator Sets* (ANSI)

(continued)

Table A.4 Continued

NECA 407-2002	*Recommended Practice for Installing and Maintaining Panelboards* (ANSI)
NECA 408-2002	*Recommended Practice for Installing and Maintaining Busways* (ANSI)
NECA 409-2002	*Recommended Practice for Installing and Maintaining Dry-Type Transformers* (ANSI)
NECA 410-2005	*Recommended Practice for Installing and Maintaining Liquid-Filled Transformers* (ANSI)
NECA 420-2007	*Standard for Fuse Applications* (ANSI)
NECA 430-2006	*Standard for Installing Medium-Voltage Metal-Clad Switchgear* (ANSI)
NECA/IESNA 500-2006	*Recommended Practice for Installing Indoor Commercial Lighting Systems* (ANSI)
NECA/IESNA 501-2006	*Recommended Practice for Installing Exterior Lighting Systems* (ANSI)
NECA/IESNA 502-2006	*Recommended Practice for Installing Industrial Lighting Systems* (ANSI)
NECA 503-2005	*Standard for Installing Fiber Optic Lighting Systems*
NECA/BICSI 568-2006	*Standard for Installing Commercial Building Telecommunication Cabling* (ANSI)
NECA 600-2003	*Recommended Practice for Installing and Maintaining Medium-Voltage Cable* (ANSI)
NECA/NEMA 605-2005	*Recommended Practice for Installing Nonmetallic Underground Utility Duct* (ANSI)

Index

A

abandoned low-voltage cables, 122
accepted industry practices, 23. *See also* standards outside *NEC*
accessible, defined, 32
accredited standards developers (ASDs), 126
accumulated wires and cables, 68–72
adequacy of temporary power systems, 105
adjustment factors for conductor ampacities, 55
adoption of *NEC*, 16–17
AFCI (arc-fault circuit interrupter), defined, 44
agricultural buildings, bonding, 84
AHJ (authority having jurisdiction), 17–21
  approval of electrical installations, 21–23
  defined, 33–34
air-conditioning equipment, 92
alarm circuits, 115–117
  fire alarm code, NFPA, 133–134
aluminum conductors, 48
amendments to *NEC* rules, 151–153
American National Standard, 151
ampacities, conductors, 54–55
  overcurrent protection, 55–57
ampacity, defined, 32, 60
annexes of *NEC*
  about, 6–7
  Annex C, 6, 71–72
  Annex D, 7
  Annex H, 7
  list of, 5
ANSI standards, 151
  ANSI/NFPA standards, 127
  ANSI/UL standards, 24–25
approval
  defined, 32–33
  of electrical installations, 21–23
  of new standards, 151
  responsibility for. *See* AHJ (authority having jurisdiction)
arc-fault circuit interrupter, defined, 44
arc flash protection, 77
arcing equipment, 99

area lighting, 16
Article 90, 6
Article 100 term definitions, 31
articles of *NEC*, 8, 148–149. *See also* numbering system of *NEC*
  numbering system of, 7, 68–70
artificial lighting of electrical equipment, 79
ASDs (accredited standards developers), 126
assembly occupancies, 95
associations organizations on code-making panels, 149
audience areas (theaters), 95
audio systems, 95, 120
authority having jurisdiction (AHJ), 17–21
  approval of electrical installations, 21–23
  defined, 33–34
automotive vehicles. *See* recreational vehicles
AWG (American Wire Gauge), 48

B

ballasts, 89
batteries, 93
battery-powered DC systems, 119
boatyards, 79, 103–104
bodies of water. *See* wet locations
bonding, 65, 83–84
  defined, 34, 84
bonding jumper, defined, 34
boxes. *See* enclosures
branch-circuit conductors, 50
  defined, 34
  lighting and appliance branch-circuit panelboards, 92
  load calculations, 60–61
  overcurrent protection, 55–60, 87
  permissible number of, 81
  voltage drops on, 63
breakers
  mounting heights of, 79
  overcurrent protection, 81
  as switches, 79, 83

Credits

Figure 2.4
Copyright 2007, Underwriters Laboratories Inc.

Figure 2.5
Copyright 2007, Underwriters Laboratories Inc.

Fig 2.6
Copyright 2007, Underwriters Laboratories Inc.

Fig 2.7
Copyright 2007, Underwriters Laboratories Inc.

Figure 2.8
© European Communities, 1995–2007.

Fig 2.15
Courtesy of AFC Cable Systems, Inc.

Fig 3.4
Courtesy of Square D Company.

Fig 3.7
Courtesy of Arlington Industries, Inc.

Fig 4.4
Courtesy of Square D Company.

Fig 4.8
Modified from Thomas Lighting, a division of Genlyte Thomas Group LLC.

Fig 4.15
Courtesy of Hubbell Incorporated.

Fig 4.17
Used with permission, Acuity Brands Lighting, Inc.

Fig 5.5
Courtesy of Pepperl + Fuchs

Fig 5.6
Courtesy of Arlington Industries, Inc.

NEC boxes are verbatim excerpts from NFPA 70®-2008, *National Electrical Code.*

NEC and National Electrical Code are registered trademarks of the National Fire Protection Association, Quincy, MA.

NFPA 70E is a registered trademark of the National Fire Protection Association, Quincy, MA.